Bell Boeing V-22 Osprey

Tiltrotor Tactical Transport

Bill Norton

AeroFax
An imprint of
Ian Allan Publishing

Bell Boeing V-22 Osprey

ISBN 1 85780 165 2

Published by Midland Publishing
4 Watling Drive, Hinckley, LE10 3EY, England
Tel: 01455 254 490 Fax: 01455 254 495
E-mail: midlandbooks@compuserve.com

Midland Publishing and Aerofax are imprints of
Ian Allan Publishing Ltd

Worldwide distribution (except North America):
Midland Counties Publications
4 Watling Drive, Hinckley, LE10 3EY, England
Telephone: 01455 254 450 Fax: 01455 233 737
E-mail: midlandbooks@compuserve.com
www.midlandcountiessuperstore.com

North American trade distribution:
Specialty Press Publishers & Wholesalers Inc.
39966 Grand Avenue, North Branch, MN 55056
Tel: 651 277 1400 Fax: 651 277 1203
Toll free telephone: 800 895 4585
www.specialtypress.com

Design concept and layout

Edited by Jay Miller

Printed in England by
Ian Allan Printing Ltd
Riverdene Business Park, Molesey Road,
Hersham, Surrey, KT12 4RG

Contents

Introduction 3
Abbreviations and Designations 4

Chapters
1 Origination 5
2 Background 13
3 Birth and Hiatus 23
4 Starting Again 51
5 Production and Service 75
6 The Future 81
7 Osprey Described 89
8 V-22 Specifications & Performance. 110
9 Tiltrotor Gallery 112

Title page: **The first V-22, BuNo 163911 during the course of its second flight at Bell Helicopter Textron's Arlington Municipal Airport Facility in Texas.** Jay Miller

Facing page: **The deck crew of the USS *Iwo Jima* have positioned aircraft 10 (BuNo 164942) with a tug preparatory to a test flight in January of 2003. Note unique tail markings for this EMD aircraft.** NAVAIR

Below: **Fresh from Boeing's Ridley Township, Pennsylvania production facility, a fuselage for a new V-22, BuNo 165943 (aircraft 44), sits at Bell Helicopter Textron's V-22 facility in Amarillo, Texas during September of 2003 awaiting final assembly.** Jay Miller

Introduction

At time of writing the US Marine Corps MV-22B Osprey tiltrotor aircraft had yet to enter full-rate production or deployment. This was the status following nearly two decades of development and flight test that saw marked changes to the vehicle, vacillations in Congressional and Administration support despite steadfast USMC dedication to the machine, and three fatal accidents. Opinions about the Osprey throughout the aviation field was similarly divided between those who saw the tiltrotor as an aeronautical advance that would surely come with sacrifices in treasure and lives, and those who saw a complex, expensive and dangerous craft unsuitable for military employment. The controversy recalled the furor over the introduction of the AV-8A Harrier into USMC service decades before. Both offered highly desirable Vertical or Short Takeoff and Landing (VSTOL) enhancements to the Marines' arsenal in fulfilling their challenging mission. Yet, both aircraft represented new technology that was met with a mixture of suspicion and the need to learn the best means of operation and employment.

Although the history of the Osprey may be short – provided it continues into full-rate production and service – the story of its lineage, development, flight test, and characteristics provide more than enough fascinating aviation history to fill this volume. A review of tiltrotor flight technology offers a view into one of the few VSTOL approaches to be taken so far along the development path. The technology is not new, and a look at the predecessor vehicles reveals the long development path leading to the Osprey. The struggle to realize a military application through multiple programs and conceptual designs emphasizes the vagaries of the US Department of Defense weapon system acquisition process. The development of the V-22 itself, marked by drawn-out schedules, up-and-down budgets, industry teaming, crises and triumphs, is remarkable in that it produced a vehicle of such capability. So potentially significant is the Osprey in military aviation that it has spawned the US Air Force CV-22B special operations variant, concepts of follow-on designs, experimental uninhabited air vehicles, and made possible a long-dreamed of civil tiltrotor. In short, the significant and already considerable history of the V-22 Osprey more than justifies a book on the subject.

If the history of the Osprey continues, this book can be updated in the future. Feedback, research material, and additional photographs are welcome. Contact the author at williamnorton@earthlink.net.

Acknowledgements

Many individuals gave generously of their time and collected materials to help make this book possible. At Bell Helicopter Textron this included Roy Hopkins II, Chuck Jacobus, Bob Leder, Bob McClure, and Dick Peasley. Of the Boeing Company thanks go to Phil Dunford, Doug Kinneard, Jim Jagodzinski, Bill Leonard, the other Bill Norton, and Marty Shubert. From the US Navy the author is grateful for support from Ward Carroll, Gidge Dady, and Linda Drew. Marines who assisted include Lieutenant Colonel 'Curly' Culp and Major Chris Seymour. From the US Air Force debt is owed to Lieutenant Colonel Tom Currie, Major Tom Goodnough, Major Greg Weber, and John Haire. Marty Maisel, formerly of NASA Ames, was especially helpful. Thanks also to retired Bell test pilot Ned Gilliand. Jay Miller and the archivists at the Jay Miller Collection provided indispensable help. A special thanks to the late John Schneider, formerly of Boeing Vertol, Michael Hirschberg of *Vertiflite* magazine, and Ken Katz.

Bill Norton
January 2004

Abbreviations and Designations

A	Amperage
AC	alternating current
a/c	aircraft
AEW	airborne early warning
AFB	Air Force Base
AFCS	Automatic Flight Control System
AFFTC	Air Force Flight Test Center
AFSOC	Air Force Special Operations Command
amp	amperage
AMT	Air Maneuver Transport
AOA	angle of attack
APLN	airplane mode
APU	auxiliary power unit
AR	aerial refueling
ASW	anti-submarine warfare
ATV	Air Test Vehicle
aux	auxiliary
AVSS	Active Vibration Suppression System
BFWS	blade fold/wing stow
bhp	brake horsepower
BIT	built-in test
BuNo	Bureau Number
C	Centigrade
cal	caliber
CAP	Composite Aircraft Program
CAT	cabin aux tanks
CDU/EICAS	Control Display Unit/Engine, Instruments, Crew Alerting System
CFB	Canadian Forces Base
cg	center of gravity
c/n	construction number
COD	carrier onboard delivery
COEA	Operational Effectiveness Analysis
CFG	constant frequency generators
cm	centimeter
CMDS	countermeasures dispensing system
CONV	conversion mode
CSAR	combat search and rescue
CSMU	Crash Survivable Memory Unit
DC	direct current
DMS	Digital Map System
DoD	Department of Defense
DoN	Department of the Navy
DT	development test
DU	Display Unit
EAPS	Engine Air Particle Separator
ECL	Engine Condition Lever
ECS	environmental control system
ECU	Environmental Control Unit
EMC	electromagnetic compatibility
EMD	Engineering and Manufacturing Development
EW	electronic warfare
F	Fahrenheit
FADEC	Full-Authority Digital Electronic Control
FBW	fly-by-wire
FCC	flight control computers
FD	Flight Director
FLIR	Forward-Looking Infrared
flt	flight
FM	frequency modulated
FMU	Fuel Management Unit
FOV	field of view
fpm	feet per minute
fps	feet per second
FRP	full-rate production
FS	federal standard
FSD	Full-Scale Development
ft	feet
FTR	Future Transport Rotorcraft
FY	Fiscal Year
G	acceleration due to gravity
GAO	General Accounting Office
GFE	government furnished equipment
GRDP	ground refuel/defuel panel
GTA	ground test article
GW	gross weight
helo	helicopter
HF	high frequency
HIFR	Hover In-Flight Refueling
hp	horsepower
HROD	high rate of descent
hrs	hours
HSD	Horizontal Situation Display
HSX	Helicopter Sea eXperimental
HUD	head-up display
H-V	height-velocity
H/WOG	Hoist/Winch Operator's Grip
HX	Helicopter eXperimental
HXM	Helicopter eXperimental Marines
Hz	Hertz
ICDS	interconnected drive shaft
ICS	inter-communication system
IFF	Identification Friend or Foe
IGE	in ground effect
IMC	instrument meteorological conditions
in	inches
IOC	initial operational capability
IOT&E	Initial Operational Test and Evaluation
IPS	Ice Protection System
IR	infrared
IRS	Infrared Suppressor
IT	integrated testing
ITT	Integrated Test Team
JSOR	Joint Services Operational Requirements
JTAG	Joint Technology Assessment Group
JVX	Joint services advanced Vertical lift aircraft (eXperimental)
kg	kilogram
km	kilometer
kts	knots
kVA	kilovolt-amps
kW	kilo-Watts
lb	pounds
LCD	liquid crystal display
LDS	Laser Detector Set
LHA	Amphibious Assault Ship (General Purpose)
LHD	Amphibious Assault Ship (Multi-purpose)
LPD	Amphibious Transports Dock
LPI	low probability of intercept
LRIP	low-rate initial production
LSD	Dock Landing Ships
lt	liters
LTM	Lateral Translation Mode
LWINS	Light Weight Inertial Navigation System
LZ	landing zone
m	meters
MATT	Multi-mission Advanced Tactical Terminal
MAW	Marine Air Wing
max	maximum
MC	Mission Computers
MCAS	Marine Corps Air Station
MDL	mission data loader
MFD	multi-function display
MFS	Manned Flight Simulator
mi	statute mile
min	minimum
MLR	Medium-Lift Replacement
mm	millimeter
MMR	multi-mode radar
MOTT	Multi-service Operational Test Team
mps	meters per second
MTE	Modern Technology Demonstrator Engine
MWGB	midwing gearbox
MWS	Missile Warning System
NAS	Naval Air Station
NASA	National Aeronautics and Space Administration
nav	navigation
NAVAIR	Naval Air Systems Command
NBC	nuclear, biological and chemical
nm	nautical mile
No	number
NOE	Nap-of-the-Earth
NORM	normal
Nr	rotor speed
NVG	night vision goggles
OAT	outside air temperature
OBOGS	On-Board Oxygen Generating System
OGE	out of ground effect
OLC	Opposed Lateral Cyclic
OPEVAL	operational evaluation
OT	operational test
Pax	Patuxent River NAS
PFCS	primary flight control system
PRGB	proprotor gearbox
PRTV	Production Representative Test Vehicles
psf	pounds per square foot
psi	pounds per square inch
PU/SS	pitch-up with sideslip
P&W	Pratt & Whitney
P^3I	preplanned product improvement
QTR	Quad TiltRotor
RADALT	radar altimeter
RAF	Royal Air Force
RAM	reliability, availability and maintainability
rev	revolution
RFP	request for proposal
RM&A	reliability, maintainability and availability
rpm	revolutions per minute
RSI	Radar Signal Indicator
R&D	research and development
SAR	search and rescue
SATCOM	satellite communications
SCAS	Stability and Control Augmentation System
SCP	set clearance plane
SDC	shaft driven compressor
sec	seconds
SEMA	Special Electronic Mission Aircraft
SEO	single engine operating
sfc	specific fuel consumption
shp	shaft horsepower
SIRFC	Suite of Radio Frequency Countermeasures
SOCOM	Special Operations Command
SOF	special operations forces
SPECOPS	special operations
STA	static test articles
STO	short takeoff
STOL	short takeoff and landing
TA	terrain-avoidance
TAGB	tilt-axis gearbox
TCL	thrust control lever
TF	terrain following
TF/TA	terrain following/terrain-avoidance
TRRA	Tilt Rotor Research Aircraft
UAV	uninhabited air vehicle
UHF	ultra-high frequency
US	United States
USAF	United States Air Force
USCG	United States Coast Guard
USgal	US gallons
USMC	United States Marine Corps
USN	United States Navy
VAC	volts alternating current
VERTREP	vertical resupply
VFG	variable frequency generators
VHF	very high frequency
V/HXM	Helicopter eXperimental Marines
VRS	vortex ring state
VSS	vibration suppression system
VSLED	Vibration, Structural Life, and Engine Diagnostic
VSTOL	Vertical or Short Takeoff and Landing, VSTOL mode

Origination

Above: **The V-22 is much like any other large rotorcraft, although with twin lateral tandem rotors. Note the slight toe-out of the nacelles.** Ron Culp

The military helicopter's ability to takeoff and land vertically is of tremendous tactical utility, and it is an indispensable asset in modern warfare. However, the comparatively low airspeed and altitude at which the helicopter commonly flies makes it more vulnerable to enemy fire than airplanes. The helicopter is typically constrained to a service ceiling of around 10,000-20,000ft (3,050-6,100m), usually preventing it from flying above weather. In fact, the all-around performance of the helicopter is commonly less than fixed-wing, conventional takeoff and landing airplanes of similar weight. The tail rotor on single-rotor helicopters is a marked drain on engine power, adds to aircraft drag and noise, and is an ever-present hazard. The helicopter reached the practical limits of its capabilities decades ago in terms of size, speed and range. Because of the rotor aerodynamics, a practical limit of 200kts is general acknowledged for rotorwing aircraft.

One answer to the helicopter's limitations has been to combine the speed, range, endurance, payload, maneuverability, and superior survivability of the airplane with the vertical lift capabilities of a helicopter. The result is the Vertical or Short Takeoff and Landing (VSTOL) aircraft. If even a minimal runway surface is available, such an aircraft could perform a short takeoff and landing (STOL) when vertical takeoff is precluded by weight or ambient conditions. The VSTOL aircraft have safety advantages over the airplane such as eliminating or reducing high-speed ground rolls for takeoff and landing, and executing off-field emergency landings into a confined space. On the other hand, VSTOL aircraft frequently have little power margin and an engine failure while in hover or slow flight, even for a multi-engine machine, can mean an immediate descent at perhaps high sink rate. This, however, is a commonly accepted characteristic of most helicopters.

Many approaches to achieving VSTOL flight have been explored. The general design requirement is a vertical component of thrust/lift greater than the weight of the aircraft to permit vertical takeoff and hover. Normal propulsive thrust must then be available for forward flight. The conversion between vertical to forward flight must smoothly transition between the two thrust/lift generation and vectoring schemes. An adequate means of attitude control from hover, through conversion at low speed, and at cruise airspeeds is also mandatory. Helicopters do all this with an articulated rotor system and, where present, a tail rotor. Airplane flight control relies on deflecting surfaces against the passing air mass, requiring forward velocity. A VSTOL aircraft cannot use airplane controls in hover and the slow-speed end of conversion and reconversion. A rotor or some other source of adequate control power must be available.

Over decades nations and corporations have invested considerably in VSTOL. A fascinating assortment of these machines have been built and tested, employing virtually all-conceivable approaches to VSTOL flight. In the US, each armed service operates transport helicopters for aerial assault, search and rescue, and vertical replenishment. All sought the potential benefits of a VSTOL transport. Through these decades only a few military designs, the tri-service LTV-Hiller-Ryan XC-142 being one, came even close to production before being ultimately judged unsuitable. Designs to fill other combat and support missions that would benefit from VSTOL have also

met with limited success. Throughout the world only the Harrier 'jump jet' fighter-bomber, first flown in 1960, and the later Yakovlev Yak-38 naval fighter have seen production. However, only the Harrier can be considered truly successful. Admittedly, these specialized aircraft are inferior to comparable warplanes in nearly all respects save for their VSTOL capability. Yet they represent useful systems in a mix of modern air combat weapons.

There are decisive reasons why VSTOL has almost always proven disappointing. The weight and cost penalties are usually too great, resulting in expensive machines with marginal performance when compared with helicopters and fixed-wing aircraft. The large excess power required for hover has required a high thrust-to-weight ratio. The propulsion system has frequently represented a disproportionate percentage of the vehicle's empty weight for a reduction in range and payload, plus adding considerably to the machine's cost, complexity, and maintenance demands. Hover performance has generally been poor, characterized by very high fuel consumption. The conversion from vertical to forward flight and back again has also been a challenging stability and control problem, complicated by a narrow conversion corridor for some configurations. The conversion corridor is the range of acceptable thrust vector angle as a function of airspeed.

Operational problems have also been endemic to many VSTOL designs. Many are characterized by high velocity columns of air, called downwash, hitting the surface beneath the aircraft during vertical takeoff and landing. This can cause surface erosion with high-energy bits potentially striking and damaging the aircraft or nearby personnel and equipment. This air will spread out along the ground as a ground plume or ground wash as another potential hazard. Additionally, this air can rising up to 'recirculate', interacting with aircraft aerodynamics. It may spoil lift in hover (called 'suck-down') and can carry ground material aloft that can damage the airframe and engine. If the ground plume is hot, as from a vertically oriented engine exhaust, the recirculate air ingested into the engine(s) ('reingestion') will produce a reduction in thrust. The high temperatures can also have detrimental effects on other aircraft components. The generation of high velocity air is commonly accompanied by very high noise levels; annoying and possibly hazardous to personnel and aircraft structure given long exposure.

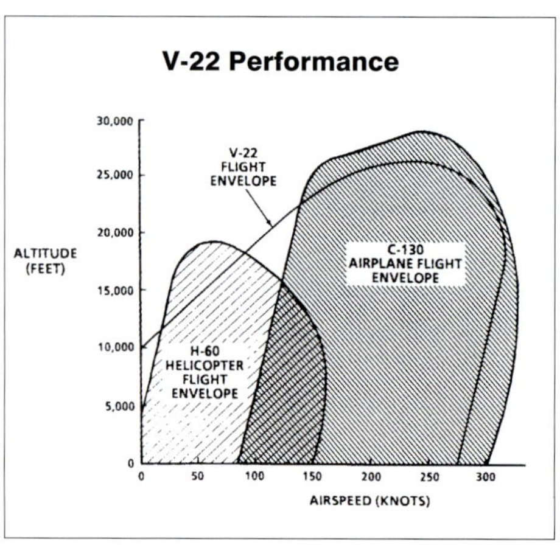

Achieving VSTOL is a matter of engineering and performance tradeoffs, and the tiltrotor generally sacrifices less for its benefit than other VSTOL concepts. The best use of the tiltrotor has generally appeared to be as a medium-lift transport where moderately high cruise airspeeds are required, yet also needing to make several stops with brief low speed and hovering operations.

The Tiltrotor

Most tiltrotor designs have the proprotors and engines together in rotating wingtip nacelles. The basic scheme is that the aircraft takes off as a helicopter (referred to as 'helicopter mode' or VSTOL) with the two rotors/nacelles vertical or 90°. These are then rotated forward to 0° for conversion to high-speed wing-borne flight ('airplane mode', APLN). Hence, the 'proprotor' blades and hub serve dual use as helicopter rotors and airplane propellers. The counter-rotating proprotors on either side of the fuselage naturally cancel the opposing rotor torques to eliminate the tail rotor. Hover performance is not as great as a helicopter with its larger rotor diameter, but this sacrifice is accepted for the comparatively high cruise airspeed. For APLN, flight control surfaces on the wing and tail take effect as airspeed increases following conversion.

The maximum speed of the tiltrotor is much greater than a comparable helicopter, and with similarly improved endurance. The tradeoff is typically slower cruise speeds than an airplane of comparable weight and power. In VSTOL the airspeed upper limits are still defined by rotor overstress and retreating blade stall. In APLN the low speed limits are set by wing stall, with the propeller wash over the wing helping to reduce this speed. The prop slipstream also helps to ensure adequate flow into the engine intake even at extreme attitudes and low airspeed.

Above: **This montage shows the tiltrotor concept from helicopter mode for takeoff and landing at the right side, conversion to forward flight in the middle with forward tilting of the twin proprotors, and high-speed airplane mode at the left with the proprotors serving as propellers.** Bell Helicopter

Below left: **This generalized diagram compares the V-22's speed-altitude flight envelope with that of a common tactical transport helicopter and airplane, the Sikorsky H-60 and Lockheed C-130. The tiltrotor nicely encompasses helicopter and airplane capabilities.** Author's collection

Below right: **A short takeoff (STO) has the nacelles at about 60° and the pilots rotate at the appropriate airspeed.** NAVAIR

Above: **A modest forward tilt of the nacelles and a short ground roll allows takeoff at gross weights or ambient conditions that would preclude a vertical takeoff.** Ron Culp

Below right: **The value of the level deck angle possible with the tiltrotor during transition to forward flight is graphically illustrated here. Aircraft 10, during its 'return to flight' on 29 May 2003, converts effortlessly while the SH-60 safety chase helicopter beyond assumes a marked nose-down attitude to keep up – the rotors of the two aircraft nearly parallel.** Navy

The cockpit controls serve common functions regardless of flight mode. In hover and low speed flight, with the nacelles tilted near vertical, the collective (or thrust control lever, TCL) and cyclic (stick) provided familiar helicopter functions, and the proprotors employ helicopter control mechanization. Lateral cyclic for roll and translation (sideward flight) commands change in proprotor blade pitch angles as they come around in rotation. This produces either a sideways tilting of the rotor disks due to asymmetrical proprotor lift, or differential collective pitch (uniform but opposed blade angle change on each proprotor for differential lift), or a mix of both side-to-side. Pitch control from longitudinal cyclic displacement gives fore and aft tilting of the rotor disks. For rearward translation, aft cyclic also brings the elevator up to keep the tail from dropping due to airflow producing a down tail load. In either axis the cyclic produces increasing rate depending upon the magnitude of displacement. Directional (yaw) control with pedals uses differential cyclic pitch – the rotor disks tilting differentially forward and aft to produce a flat rotation about the vertical axis. In VSTOL the TCL commands proprotor collective (uniform) pitch and engine power simultaneously for disk lift variation to change altitude or hover, but power only in APLN. The TCL commands symmetric rotor or mast torque in both VSTOL and APLN.

As the aircraft accelerates through high-speed conversion, the controls change their functions and the pilot's control strategy has to progressively change to resemble that of a conventional fixed-wing aircraft. In APLN the rudder pedals produce yaw while the stick becomes a climb, dive and roll rate controller, moving the ailerons/flaperons and an elevator. The TCL input (power) is used as a simple throttle to set the longitudinal thrust while longitudinal stick is used to manage the aircraft energy state by increasing or decreasing the flight path angle at relatively constant speed, or allowing the aircraft to accelerate and decelerate.

In some areas of the airspeed and nacelle angle conversion corridor the choice of control technique can be uncertain. The 30° nacelle setting at the lower end of the acceptable airspeed for that angle is one such ambiguous condition in then V-22, and accompanied by airframe buffet. However, pilots are trained to use a few nacelle settings and certain airspeeds during transition to help avoid confusion. The tiltrotor normally spends little time in transition.

The conversion is begun at an airspeed at which the wing is gaining in lift as the rotor lift decreases with tilt angle. This airspeed must also be such that the wing and tail control surfaces are sufficiently effective to control the air-

Both pages: **This series of drawings and notes illustrates how the tiltrotor is controlled in flight through the pilot thrust control lever, cyclic (stick) and directional pedals. A. Thrust Control (power); B. Forward Cyclic; C. Aft Cyclic; D. Lateral Cyclic (right); E. Pedal (left)**
Bell Helicopter

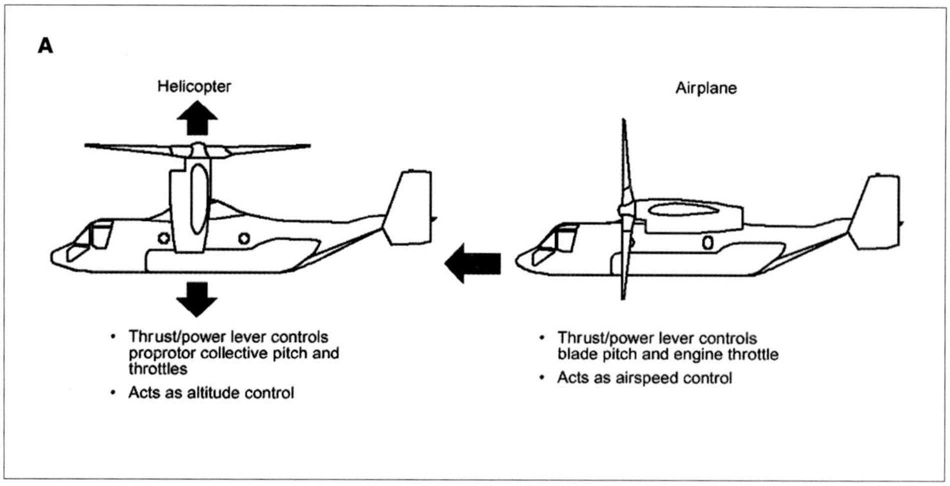

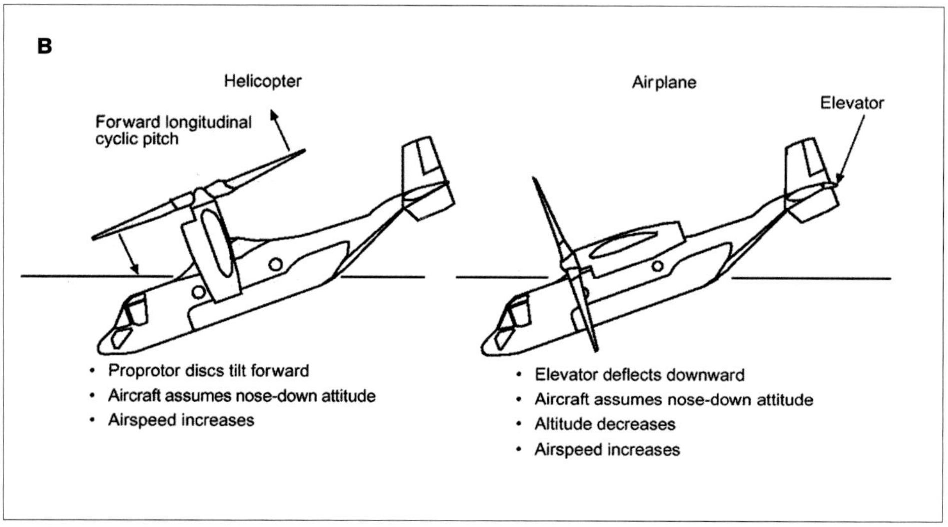

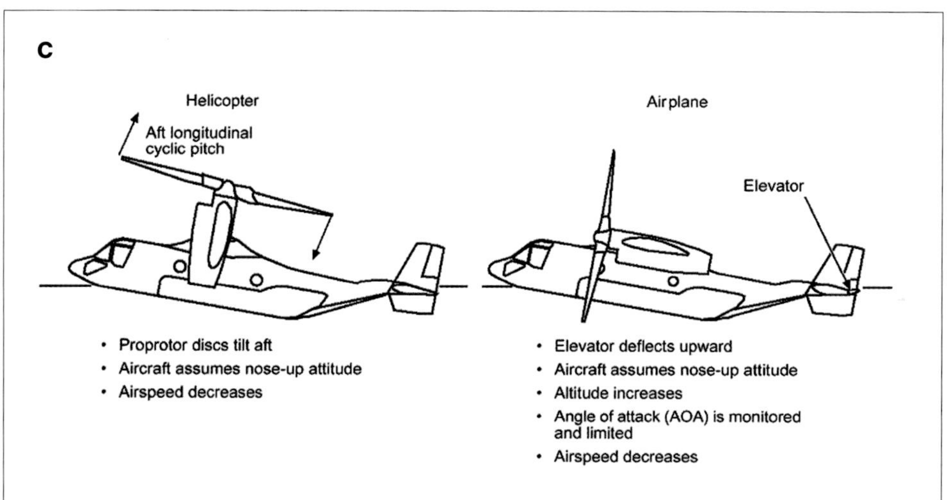

craft as the proprotor hub controls become ineffective as tilt angle decreases. Intermediate proprotor positions ('conversion mode', CONV) allows for very short rolling takeoffs and landings with a greater payload than for a vertical takeoff, provided ground clearance for the rotors is maintained. The system can also provide advantageous thrust vectoring 'up and away' for enhanced maneuverability.

With the proprotors placed far outboard of the fuselage, excessive asymmetrical rotor lift or propeller thrust would generate rolling or yawing forces too great for conventional rotor hubs or airplane flight control surfaces to overcome. This means it is best that a transmission interconnect drive shaft through the wing join the engines, or at least the proprotors. Likewise, the rotors or engines/rotor combination must tilt in precise unison to avoid loss of control. This has usually required an additional tilt axis cross shaft or fail-safe electronic control.

In APLN the proprotors rotated up on the inboard side, generating an air swirl opposite the wingtip vortices. These vortices, common to any wing, are from high-pressure air on the bottom flowing up around the wingtip to the low-pressure region atop the wing and generate drag. The proprotors counter of this flow for reduced drag.

The proprotor blades must be designed to operate efficiently as helicopter rotors and airplane airscrew. This challenging requirement has yielded short and broad blades with considerable twist – 47.5° on the V-22 versus 8° for a typical helicopter. A measure of this is disk loading, or aircraft weight divided by the area of the circle(s) formed by the rotor diameter. A typical disk loading for the medium lift V-22 is 20psf (99kg/m^2) versus a typical 6psf (27kg/m^2) for the comparable CH-46 and 10psf (50kg/m^2) for the heavy lift CH-53D. The helicopter's low disk loading is more efficient in hover and generates a comparatively mild downwash. High disk loading means more power required to lift the same aircraft, equating to more weight and high fuel consumption in hover, and greater downwash velocity. Consequently, the tiltrotor normally requires more power for hover, translation and forward flight in VSTOL than a helicopter, but less than many other VSTOL designs. Plus, the tiltrotor possesses greater rotor drag in edge-wise flight. However, high disk loading is preferable for cruise flight with the blades working as propellers. Here, too, the design is not ideal and the rotation rate must be reduced in APLN for improved proprotor efficiency. However, tiltrotor hover efficiency is much better than almost all other VSTOL designs. The presence of the wingtip nacelles also contributes considerable drag during translation and conversion. So, a helicopter performs better than the tiltrotor in hover and translation, but cannot fly as fast in cruise. The airplane performs better in cruise, but cannot takeoff vertically. The tiltrotor is the epitome of engineering compromise.

A long-standing challenge in tiltrotor design has been avoiding rotor, pylon (combined tilting nacelle and power transmission gearbox), and structural instabilities. The elastic responses of the structure and rotor dynamics can interact with the aerodynamic forces to produce structural oscillations that can grow to destructive magnitude. All these factors change with flight condition, fuel weight in the wing, nacelle angle, blade flapping, and rpm, making for a complex design problem. The tiltrotor wing is typically thick for the purpose of ensuring suitable stiffness and aeroelastic stability. Yet, the airframe and rotor system must be lightweight.

The tiltrotor offers some unique VSTOL advantages that support its claim to being revolutionary. The principal benefit is that the engines and thrust generation devices for vertical, STOL, conversion, and cruise flight are the same. It combines well understood helicopter and airplane technologies. The tiltrotor usually has a more generous conversion corridor. The high-speed end of the corridor is defined by prohibitive rotor and nacelle/wing interface loads. The low-speed boundary is usually determined by wing stall. The tiltrotor is generally easier to stabilize than other configurations, especially during conversion and reconversion where both the helicopter and airplane controls are available to greater or lesser degrees as airspeed changes. Turn performance across its speed range is superior to the helicopter.

The airplane configuration allows the tiltrotor to be flown to altitudes far above that of a helicopter, or more comparable with turboprop aircraft. This allows flight above weather whereas a helicopter would be grounded, forced to divert, or fly in adverse conditions under the weather. The exterior noise during hover and transition is about that of a heavy helicopter and much less noisy than nearly all other VSTOL designs. In APLN the tiltrotor is quieter than a turboprop aircraft by virtue of its lower proprotor tip speed, and only a three quarters the level of a helicopter – and without the distinctive 'whop' – enhancing military covertness. Vibration levels in APLN, where the aircraft spends most of its time, is substantially less than in VSTOL, reducing component wear and failure rates.

In the conversion to forward flight, flight at intermediate nacelle angles, and approach to hover the tiltrotor's deck angle can be maintained level or at a nose-low attitude for improved visibility. This is accomplished with thrust vectoring independent of aircraft attitude by using cyclic opposite the proprotor tilt angle (for example, aft stick for a forward tilt). The helicopter must raise its nose dramatically to rapidly bleed speed in the approach to hover. This is where the windows at the helicopter pilot's feet become most important in maintaining sight of the landing zone. The same is true for the acceleration to forward flight. The helicopter must point its nose down, sometimes considerably, to affect a rapid acceleration. The proprotor tilt also helps to make upslope and downslope landings safer. The rotor disk can be kept level to maximize longitudinal cyclic authority to handle gusts and unexpected disturbances, while the deck angle is made to match the slope.

The principal penalties of the tiltrotor are found in the weight and complexity of added gearboxes, tilt mechanisms, and cross shafting, all contributing to increased unit and support costs. The issue of the rotor downwash or download on the wing and fuselage in helicopter mode is an endemic tiltrotor concern and one of the primary hurdles to achieving good hover performance. A reduction of just 1% in download can add 500 lb (225kg) of payload. The outflow of the rotors impinging on the upper wing surface underlying the rotors is the opposite of that desired for lift, robbing the aircraft of potential payload capacity. The flow on the wing moves inboard, meeting and fountaining up at the center to be recirculated through the rotors for a loss in rotor lift. These effects can be reduced with deployed flaps, wing fences, and wing design choices. The leading edge-to-trailing edge directions of proprotor rotation also reduces download. As the aircraft begins to move forward the rotor downwash is 'blown' aft such that as little as 20kts is required to substantially reduce download.

Fuel, hydraulics, and electrical connections must pass through the rotating nacelle interface that is an added maintenance burden potentially impacting overall system reliability. The wingtip nacelles increase aircraft roll and yaw inertia, requiring more control power for some maneuvers in all flight modes. Conversely, with the proprotors far removed from the fuselage, considerable control power is available in VSTOL. Placing the engine far outboard somewhat reduces the risk to occupants from engine fires and turbine bursts.

When close to the ground, the ground wash is comparable to that of a heavy lift helicopter. When close to the ground the meeting of the outwash from the opposite rotors under the centerline of the aircraft fountain up to impinge on the fuselage bottom and add to lift force. This flow is also directed forward and aft of the aircraft, and can lift dust and other material to obscure vision, although peripheral vision remains good. The ground plume can recirculate and produce some loss of performance, although the high engine inlets on tilted nacelles may reduce this effect. The jet exhaust directly impinging on the ground raises surface and aircraft lower extremities heating concerns. However, the heating is much less than other VSTOL concepts and, combined with the comparatively low velocity proprotor downwash, represents an acceptably 'soft footprint'. Furthermore, extended hovers at greater height than comparable helicopters can help ameliorate such concerns.

One of the greatest flight safety concerns with the tiltrotor has been engine-out landing. For a single engine failure in a twin-engine tilt-

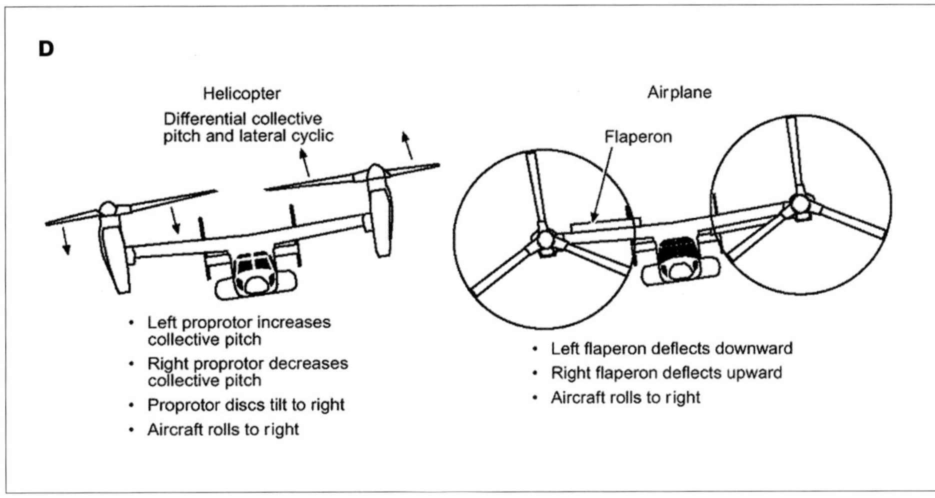

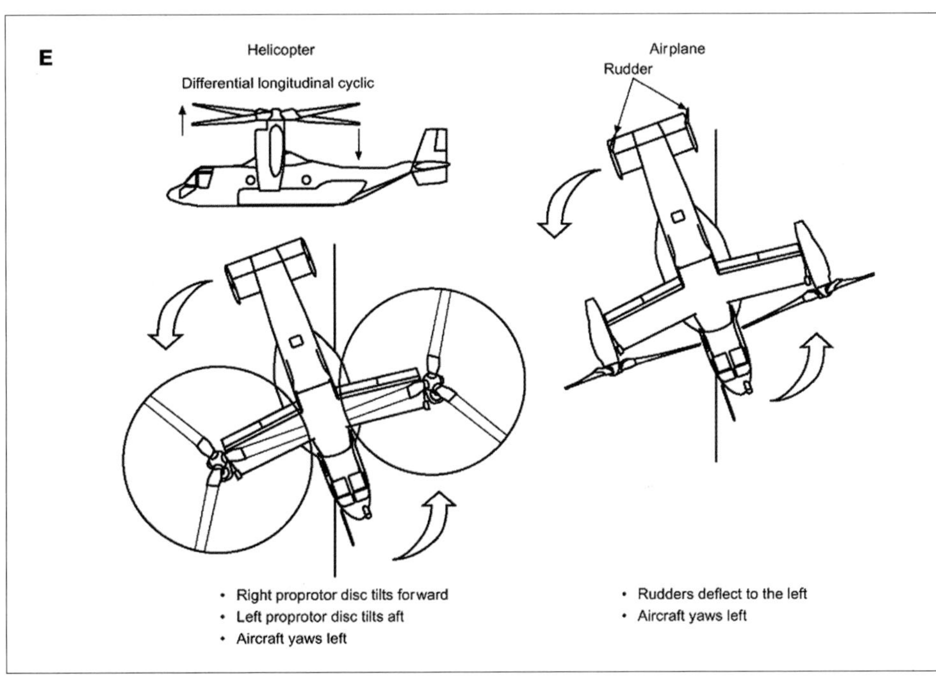

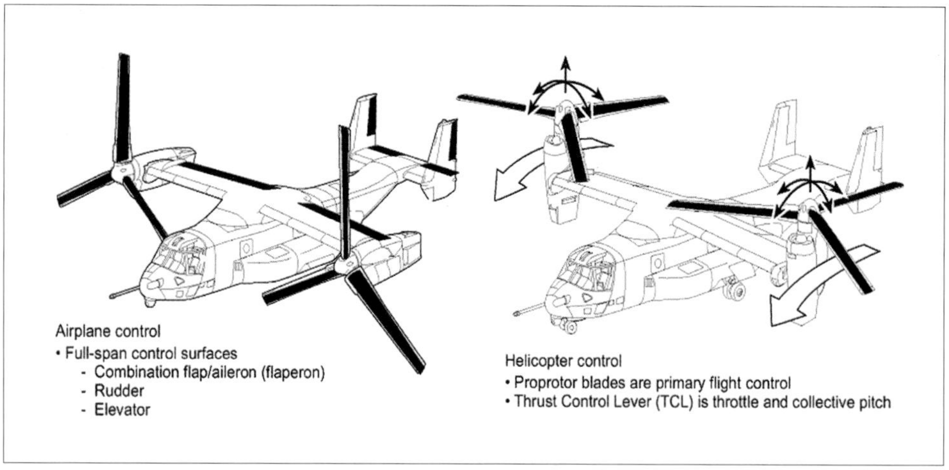

Left: **The tiltrotor flight control effectors are shown here, with both helicopter and airplane elements being used throughout the envelope to various degrees.** Bell Helicopter

Bottom: **This diagram shows the positive (dark arrows) and negative (light arrows) lift factors for in-ground-effect hover. Shown is the fountain recirculation from the center of the wing, the fountain lift at the bottom, and wing download, the minor exhaust thrust, and the groundwash.** Author

rotor, the hazard is no more severe than with any other large multiengine rotorcraft, and the V-22 has comparatively good single engine operating (SEO) performance under most conditions. Where conditions such as gross weight (GW) and ambient temperatures present sufficient thrust margin to hold a hover, a common SEO landing can be made. If performance offers little or no hover capability, a forced landing must be executed as it would for a helicopter. However, the wing provides an advantageous glide ratio (ground distance covered for altitude lost) making a roll-on short landing practical if a runway or reasonably smooth surface is within reach from a safe starting altitude. The V-22's advertised glide ratio is 9:1, but a more realistic number under operational conditions would be 4:1. This is far better than helicopters and gives a greater margin for finding a favorable landing site. A vertical landing can be accomplished, but a high sink rate may be present when approaching the ground. However, rolling the proprotors back just prior to landing may arrest forward velocity and sink rate to zero if done properly. Also, unlike a helicopter that must come in at an extreme pitch attitude to arrest speed, the tiltrotor's deck angle can be level or nearly so, increasing the likelihood of a damage-free forced landing.

For a complete power failure, the tiltrotor offers the advantage of selecting either a vertical or horizontal landing. However, each has significant limitations over the helicopter or airplane, respectively. For a horizontal landing, the approach and landing performance with the small, highly loaded wing would not compare with the engine-out characteristics of a comparable airplane. Also, the rotors must be brought up enough to prevent a proprotor strike. However, the V-22 composite blades would simply delaminate to 'broomstraw' if striking the ground, reducing the hazard of high-energy debris. The aircraft would still likely suffer substantial damage, but personnel injury would be minimized. For a vertical landing from airplane mode, a power-off reconversion must be executed prior to the forced landing. The proprotors will continue to turn because of the airflow through the rotor (autorotation).

There is the danger of system failures preventing tilt of the proprotors from airplane to helicopter mode, or at least moving the rotor tips above the ground plane for a safe STOL landing. This risk is sufficiently reduced with tilt axis system redundancy and backup power. Furthermore, during reconversion the unpowered proprotors pass through a point where there is insufficient flow to maintain rotational speed. This point (70° tilt for the V-22) must be transitioned quickly while energy remains in the rotating system by using the maximum tilt rate. Also, maintaining rotation of the proprotors and interconnecting drive shaft near the normal rpm range is usually essential in energizing the electrical or hydraulics system that powers the conversion actuators and other essential functions. Should the proprotors stop turning prior to reaching helicopter modes – a remote possibility unless the pilot makes a profound error – the aircraft will come down quite fast and have to execute a landing like an unpowered airplane, possibly with proprotor blades below the bottom of the aircraft.

With the proprotors in helicopter mode during an unpowered approach and landing, the aircraft nose is pitched down to ensure maximum upflow through the rotors and 100% rotor speed. Flaps are set to ensure adequate wing lift. Even if all goes well, the tiltrotor has quite marginal autorotation capabilities. However, autorotation is not a principal design condition for the proprotors because of the very low probability of this failure state and the other horizontal landing option. For the V-22, the descent will be very steep and the rate initially quite high in comparison to a helicopter because of higher disk loading – 3,000-4,000fpm (15-20mps) versus 2,000-2,500fpm (10-13mps). Delaying full reconversion to the last few seconds before touchdown, and using the small amount of aft proprotor tilt usually available, will greatly arrest the landing sink rate. As with the helicopter, a cyclic flare and judicious collective application can permit a low vertical and forward velocity at landing. However, the high rate of descent will make all this a decidedly difficult proposition, with the final actions required in a split second.

The tiltrotor can theoretically be set down just as well as a helicopter executing an autorotation landing, as demonstrated in simulation. But, a STOL approach and landing combining autorotational and wing lift would be the safer option.

Another concern with the tiltrotor, or perhaps any VSTOL aircraft, is that control or propulsion system failures can immediately generate a state where a catastrophic accident will result. These would be failures producing pronounced asymmetric proprotor lift or thrust. This problem is addressed with considerable system redundancy and built-in tests.

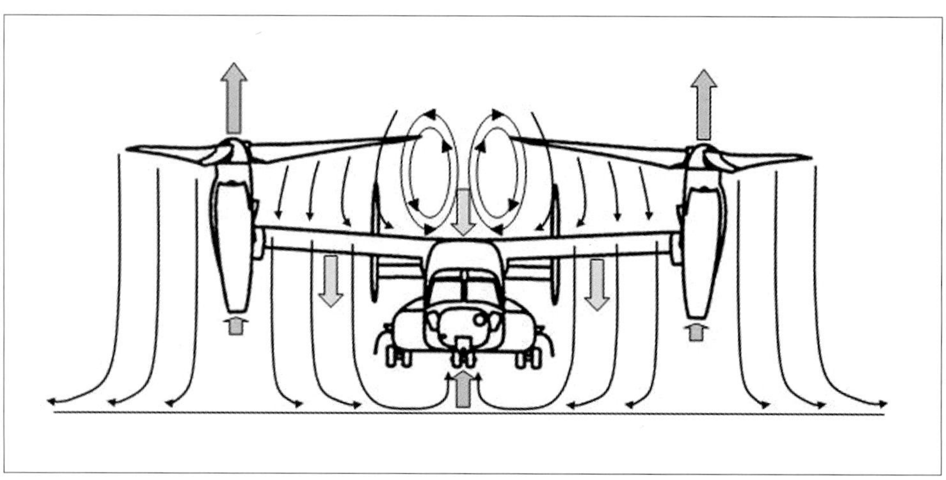

Photographs on the facing page:

Aircraft 1 during a test flight south of the Bell Helicopter facility at Arlington Municipal Airport. Noteworthy are tufts for visual observation of airflow over fuselage, wings, and engine nacelles. Jay Miller collection

Aircraft 12 is captured in conversion mode low over the ocean. Ron Culp

01

02

EXPERIMENTAL

N546A
TRANSCENDENTAL
EXPERIMENTAL

US ARMY
4147
XV-3

U.S. ARMY
4148

Chapter Two

Background

Predecessor Tiltrotors

Aircraft with tilting propellers were conceived early in the history of manned flight, and tilting rotors soon after the advent of helicopters. Much research worldwide have yielded many tiltrotor concepts and even construction, but very few have actually taken flight. The development presented engineering challenges requiring decades of research and technology maturation, with a few experimental aircraft built to collect data, before a practical vehicle could be realized.

The first true tiltrotor aircraft to fly was built by the Transcendental Aircraft Corporation. Their Model 1-G project was helped along with some Department of Defense (DoD) funding. The tiny aircraft had a maximum hover GW of just 1,750 lb (794kg) and a wingspan of 21.00ft (6.40m). The rotors were powered by a single reciprocating engine within the fuselage. The first hover flight occurred on 15 June 1954. The machine made its first partial transition five months later, eventually flying with about 70° of rotor tilt. After over 100 flights and 23 flight hours the aircraft crashed on 20 July 1955 before completing a full conversion. The friction lock on the collective slipped, generating a steep dive and could not be recovered in time. The similarly small, 2,249-lb (1,020-kg) GW, Model 2 Convertiplane was flown in the latter half of 1956 with a more powerful engine. It is believed this machine never achieved complete conversion before the program was abandoned in 1957.

Bell Aircraft, on its way to becoming a premier helicopter manufacturer, performed tiltrotor design studies as early as the 1940s. One of their earliest concepts was the Model 50 Convert-O-Plane followed by the D-79 'rotor plane' for a single occupant. A series of tiltrotor designs followed, including the 23,100-lb (10,478-kg) D82A transport and 82B rescue 'Rotorplane', and the single-engine D-100 research vehicle. The D-118 Convertiplane was a single-engine machine with two occupants and a reconnaissance/observation role. These studies clearly indicated that such an aircraft was feasible and could yield great utility.

Bell was in an advantageous position when the US Air Force and Army announced the Convertible Aircraft Program competition in 1950. Reviews had suggested that the technology to achieve VSTOL flight was within reach. The program provided funds to resolve some of the more daunting engineering uncertainties, and eventually produce demonstration aircraft. Three proposals were selected in 1951, including Bell's XV-3 Convertiplane. Acceptance of the company's preliminary design followed with funding for construction and testing of two vehicles.

The XV-3 accommodated two pilots in tandem seating and two litter-borne casualties in a small cabin. The test aircraft was small at only 30.33-ft (9.25-m) length and 31.33-ft (9.55-m) wingspan. Its normal operating weight was just 4,800 lb (2,177kg). The 450 bhp (335kW) radial engine was mounted in the fuselage and drove twin three-bladed rotors. The power was transferred to the rotors via a short shaft to a transmission gearbox and then to the wingtips via shafts within the wing. The main transmission had a two-speed gear reduction feature for rotor speeds to be stepped down from VSTOL rpm for APLN. The shafts drove wingtip-mounted tiltable transmissions that turned the rotor masts. The design tools of the period yielded blades of comparatively low efficiency in both the VSTOL and APLN, and the aircraft soon proved quite under-powered. In fact, the aircraft could not hover out of ground effect (OGE).

The helicopter collective side lever with throttle twist-grip was retained for use in both VSTOL and APLN. The change from helicopter to airplane control of the rotor hub during conversion was entirely mechanical. Rotor tilt was commanded via a 'beep' switch on the collective that operated an electric motor and actuator located in each wingtip pod. The actuators were mechanically interconnected via cross shafting through the wing to ensure synchronized motion even in the event of the failure of one actuator. The proprotors could be tilted through their full range in just 10 seconds or stopped at any intermediate angle.

Months of ground tests at Bell's Hurst, Texas, plant began in early 1955 and included securing the aircraft to an elevated platform or 'runstand' where full tilting of the powered rotors could be performed. Ship 1 executed its first hover on 11 August 1955. Control instability during air taxi on the 18th resulted in a hard landing and minor damage. This began a long struggle with control and aeroelastic instabilities that saw repeated changes to the aircraft and long periods of ground testing. Changes included adjustments to the rotor mast length,

Facing page, top left: **The tiny Transcendental Model 1-G is shown in hover during 1954 or 1955. The two-speed reduction gearbox is visible at the wing root and above the 160 bhp Lycoming O-290-A reciprocating engine.** John Schneider via Marty Maisel

Facing page, top right: **Transcendental's Model 2 featured 18-ft (5.5-m) diameter three-bladed rotors tilted with electrical actuators. The considerable wing area without flaps implies a considerable download.** John Schneider via Marty Maisel

Facing page, middle left: **One of Bell Helicopter's earliest concepts for a tiltrotor transport aircraft was the D82B design. This version would have featured interchangeable cargo pods, the aft fairing apparently sliding forward on the fuselage to mate-up with the cockpit section for fight without the pod.** Jay Miller Collection

Facing page, middle right: **At roll-out the XV-3 (ship 1 shown) had very clean lines, including a closely cowled engine mid-fuselage, no flaps, and three-bladed proprotors. Development testing would soon yield many changes.** Jay Miller Collection

Facing page, bottom: **Seen later in flight test over Texas, ship 2 has been given a scoop intake for engine cooling, flaps, rudder tab, and a fundamentally redesigned two-bladed rotor hub. AFFTC;** Inset: **A proposal for a tiltrotor testbed using a Cessna T-37 as basis.** Jay Miller collection

Right: **Hovering at Edwards AFB near the end of its career, the XV-3 is seen with an ventral fin for enhanced lateral-directional stability and wing struts to combat structural dynamic instability.** Tony Landis collection

Left top: **Although separated by nearly three decades, the Bell D-223 tiltrotor concept was remarkably like the Osprey. Its size and engine/nacelle layout are essentially the same, with only the cruciform tail a marked variance.** Jay Miller Collection

Left bottom: **Bell's D-246A medium transport design showed an alternate engine installation scheme in fixed nacelles. The drawing does not suggest that the proprotor pylons rotate. But the mid-span line may indicate that the entire outer wing, not just the tip, was to rotate, reducing download.** Jay Miller Collection

mast mounting, and the addition of rotor dampers. The first in-flight rotor tilt was made on 11 July 1956. However, rotor instability arose again on 25 October after reaching just 70° of tilt and 80kts. The aircraft was written off in the resulting accident that seriously injured the pilot.

A fundamental redesign of the rotor ensued, yielding a two-bladed stiff in-plane or semi-rigid configuration. Following modification, ship 2 was subject to full-scale wind tunnel testing that revealed a rotor-pylon-wing instability during conversion. Bell introduced further changes that shortened the mast and reduced rotor diameter, and the wing stiffness was increased with external struts.

The resumption of flight test demonstrated that the rotor stability problems were not over, prompting more wind tunnel and analytical work. Progress was measurable when the XV-3 performed the first complete conversion for a tiltrotor aircraft on 18 December 1958. Testing by the Air Force and National Aeronautics and Space Administration (NASA) followed. Total flight time of 125 hours in over 250 flights was logged on the two XV-3s, including 110 full conversions, with at least 11 pilots flying the aircraft.

Testing included exploring the STOL potential using small wheels installed on the aircraft's skids to permit rolling takeoffs and 'run-on' landings. The machine took off in less than 200ft (61m) at 30kts with just two-thirds of available power and the proprotors at 80°. Extrapolation suggested that the aircraft could perform a short takeoff (STO) at 50% overload under such conditions. Several pilots felt the conversion corridor was generous and transition a fairly simple task, remaining within limits without undue pilot attention. A maximum altitude of 12,300ft (3,750m), airspeed of about 115kts in level flight and 155kts in a dive, and autorotations were all demonstrated. Recirculation was definitely affecting the under-powered aircraft when hovering near the ground.

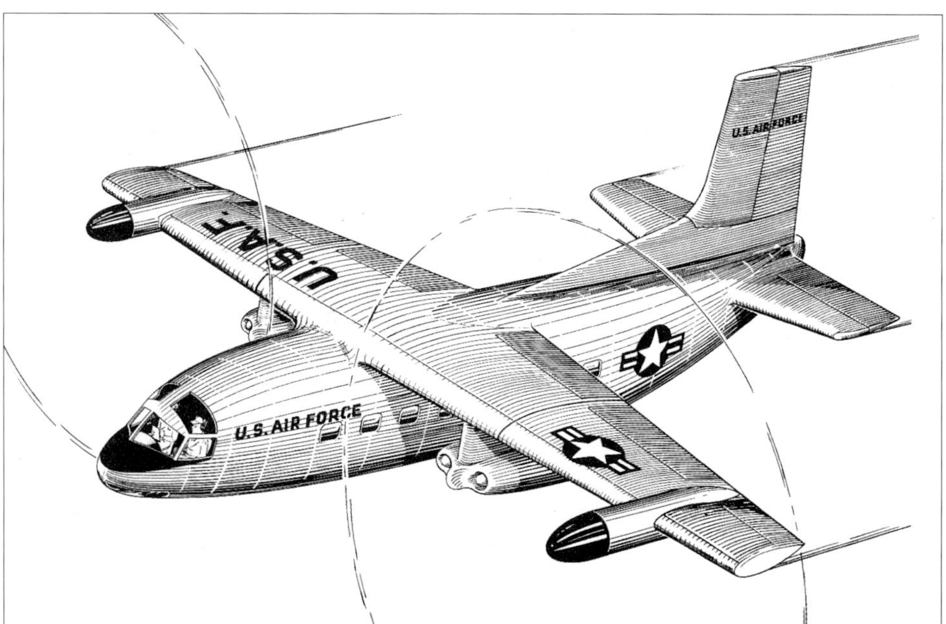

Many controllability difficulties remained plus excessive blade flapping generated by airplane maneuvering. Negative proprotor influences on lateral-directional stability required more tail area to be added. A mild longitudinal acceleration/deceleration while flying in gusty air, called 'chugging', was attributing to variations in proprotor airflow incidence angle generating uncommanded thrust and rpm changes. The rpm governor reacted to this but with a system delay that allowed the longitudinal instability to manifest itself. Many other consequences of the tiltrotor layout were revealed, providing many lessons and proving the worth of the program. It was learned that in hover near the ground the fountain of air under the fuselage was unsteady, producing an aircraft lateral darting. It was also found that when the aircraft rolled in one direction for an in ground effect (IGE) hover translation an unstable tendency to continue rolling was encountered. Consequently, pilot workload was high in some portions of the flight envelope.

Even as a proof-of-concept platform the XV-3 was only marginally successful. It proved that a tiltrotor was possible, but it did not engender confidence in its much-lauded advantages. The XV-3 could do little that a helicopter could not do better, although it was a remarkable achievement given the technology of the period. The XV-3's value as a research program was more certain. Many tiltrotor engineering challenges were overcome. It decidedly advanced the knowledge to build and operate such machines.

Despite of the mixed results of the XV-3 project, Bell Helicopter was encouraged to continue its efforts at demonstrating the tiltrotor could be practical. Rotor design efforts eventually yielded a semi-rigid rotor gimbal-mounted to the mast. The underslung hub 'floated' on a non-rotating elastomeric rubber spring for enhanced damping. This permitted large out-of-plane flapping of the rotor without individual blade flapping hinges, permitting minimum mast height with attendant stability advantages. Each blade retained a separate pitch change mechanism but the lead-lag hinges were eliminated. The hub was also designed to be compact and fit within a more aerodynamic

Right: **Another of Bell's many tiltrotor designs over the decades was the D-266 medium transport. DoD money funded considerable design research although the aircraft was never developed.** Jay Miller Collection

spinner and fairing, reducing airplane mode drag. A wing with modest forward sweep accommodated the maximum rotor flapping in APLN. This was carefully matched with a suitable pylon/wing structure that reduced potential for dynamic instability. Bell concentrated on the three-blade rotor because it provided stability benefits.

Bell design studies in response to government programs or as commercial ventures gave birth to numerous projects over almost two decades. There was the D-207 cargo convertiplane and the D-222 multi-engine tiltrotor for rescue work. The D-223 Transport Convertiplane of 1956 was actually quite similar to the V-22. It was to be a medium-size transport with rear-loading ramp. The proprotors were to be installed with two T58 turbine engines in each rotating wingtip nacelle. The advent of high power-to-weight ratio turboshaft engines was a particular boon to tiltrotor design that Bell was quick to exploit. There followed a series of convertiplane designs over the next few years including the D-224 rescue craft conceived with three turbine engines, the D-225 with a pair of Allison T56s, D-242, the 243 cargo transport, and the D-244. The 38,000-lb (17,214-kg) D-246A was quite similar to the 223 but carrying four Lycoming T55 engines, paired in fixed nacelles under the inboard wing sections. Its proprotors were at the tips of pods set inboard from the wingtips; the outer wing section tilted with the proprotors. The 32,300-lb (14,650-kg) D-252 was proposed as a tri-service transport with two T64 engines.

The D-266 tiltrotor concept was offered in response to the US Army's Composite Aircraft Program (CAP) in 1966 for which a modest research contract was awarded. The 21,000-lb (9,525-kg) Model 266 was to have two General Electric T64 turbine engines mounted either at the wing roots or in tip nacelles. The aircraft was projected to be capable of hovering at 7,000ft (2,134m) OGE, fly 350nm with an 8,000 lb (3,629kg) payload, and possess a maximum speed of 350kts. Bell and NASA tested a model of the 266 in a wind tunnel, achieving a scale speed of 478kts and producing invaluable aeroelastic data. Although CAP was canceled, the design, raised to 28,000 lb (12,700kg) gross weight, was then submitted to fulfill the USAF's Light Intra-theatre Transport requirement that was also canceled.

Boeing Vertol was performing VSTOL research involving rotor systems during this same period, with much focus on tiltwing designs. The company conducted a study of all VSTOL approaches, finding the tiltrotor the most promising. Turning their research along this line, they soon amassed a large database. In addressing the aeroelastic problem their solution was a hingeless rotor hub that had no blade flapping. By 1968 the company had conceived their Model 160 that had many similarities to the future V-22.

Bell committed to a long series of in-house tiltrotor technology development projects, sometimes with government funding, during the 1960s. In 1968 this yielded the D-267 design. The aircraft was conceived as a 9,000-lb (4,082-kg) commercial aircraft with a 750-1000-shp (560-746-kW) Pratt & Whitney (P&W) PT6 turboshaft engine in each wingtip nacelle. It was further revised as the 12,400-lb (5,625-kg) Model 300 with 1,150shp (858kW) PT6s. Bell engineering analyses and wind tunnel tests of the 300 showed promise. A 25-ft (7.6-m) diameter proprotor was tested in a tunnel. An unusual feature of the hub was collective pitch changes made via an axial shaft instead of separate swashplate actuators.

Although Bell, Boeing, other manufacturers, and government laboratories pursued tiltrotor research, major efforts had to rely upon government interest and funding. Many small programs built up engineering data and expertise required to develop the next tiltrotor aircraft.

The US Army and NASA, collaborating closely on VSTOL flight research, decided by 1971 that the time was ripe for another proof-of-concept aircraft. This would take advantage of current predictive techniques and reliable electronic stability and control technologies that promised greater ease of operation. The focus was on a concept that might eventually fill an Army air mobility requirement and serve as a commercial mid-range transport. The Army and NASA concluded that a tiltrotor was the best concept for these missions.

After a competition, Bell and Boeing Vertol were selected to perform more detailed design work. Bell began with a minimally revised Model 300 design, the 301. The final design

Bottom: **Boeing's entry in the TRRA competition was the Model 222. Innovations included wing leading edge flaps to further reduce download and fixed engines driving the hingeless proprotors mounted just inboard.** NASA via Marty Maisel

included a change to an H-configuration empennage, a GW increase, and two 1,250-shp (932-kW) Avco Lycoming T53 engines. This required a substantial change to the transmission Bell already had running on the test stand, but the existing rotor and blade design was retained. In April 1973, Bell Helicopter Textron was selected to carry on and build two XV-15 Tilt Rotor Research Aircraft (TRRA), or the XV-15.

Overall length of the XV-15 was 42.10ft (12.83m) and span between the outside rotor tips was 57.20ft (17.40m). It had a conventional airplane layout, the wing possessing 6.5° forward sweep with trailing edge flaps and outboard flaperons. The reduction in planform area in hover with lowered flaps/flaperons saved 6.5% lift that would have been lost to download. The lowered surfaces also increased lift from upwash off the ground. The aircraft was a bit over the anticipated empty weight. With this and the engine/transmission choices, the aircraft could only takeoff vertically with full fuel under optimal conditions. It had a maximum of 15,000 lb (6,804kg) GW for STOL takeoff and landing, and 14,250 lb (6,464kg) for vertical operations. The aircraft was intended to carry about 1,300 lb (590kg) of test equipment for a typical 45-minute mission. Maximum range was 435nm, although a ferry tank and auxiliary (aux) tank was eventually devised for the cabin. The longest flight was 1,475nm.

The two proprotors were connected by cross shafting in the wing. A centerwing gearbox was required because the wing dihedral and sweep mandated an angle change in the cross-shaft. With both engines operating, the shaft was unloaded but for the torque needed to turn any accessories driven off the shaft. A one-way overrunning clutch in the transmission automatically disengaged a failed engine so as not to rob the rotating system of energy. Unlike the XV-3 proprotor blades, the TRRA had broad and highly twisted blades. A helicopter-type collective or 'power lever' was installed to the left of each pilot. A power lever switch controlled rotor rpm, with governors reducing the rate for APLN.

A 'colley hat' switch on each power lever commanded tilt. Moving the switch forward commanded down rotation and aft commanded up tilt. Returning the switch to the center stopped the motion. A conversion speed switch allowed either 100% nacelle tilt rate (7.8°/sec) or 20% (1.5°/sec). The nacelles could be rotated down to 0° and up to 95°. A second switch permitted the nacelles to move at 1.5 deg/sec to discrete angles of 90°, 75°, 60°, and 0°. Tilting was accomplished with a hydraulically powered actuator pivot-mounted between each nacelle and the wing box beam at the wingtips. These actuators were interconnected for synchronized operation via a separate cross shaft and phasing gearbox in the wing to ensure uniform tilt angles. The aft 5° tilt supported aircraft backing on the ground, assisted rapid deceleration on the ground or in flight, and higher rearward flight speed than the ±12° of cyclic alone could provide. Taxi speed could be controlled via the coolie hat, with only 5° of forward tilt needed for most conditions.

A three-axis Stability and Control Augmentation System (SCAS) helped reduce pilot workload while improving aircraft response and flying qualities. However, even the SCAS and all other electronic system aids deactivated the aircraft could be flown with tolerable workload. This was a considerable achievement given that many non-augmented VSTOL designs had suffered from grievous controllability difficulties.

Of the characteristics experienced with the XV-3, some were inherent to the tiltrotor design and inevitably exhibited by the XV-15. These included a nose-down pitching during transition, large variations in power-required during the approach to hover, and lateral instability during IGE hover. They are characteristics of the laterally displaced rotors, the under-lying wing, and rotor wash interacting with the horizontal stabilizer. As the tiltrotor climbs and accelerates during nacelle rotation, the rotor IGE lift addition diminishes as wing lift increases. The latter effect is seldom adequate to prevent a momentary decrease in overall lift that requires the pilot to bring up power to prevent a slight sinking. This effect was of less significance with the XV-15 because it had more excess power than the XV-3. Like the XV-3, the TRRA displayed stick force reversal at low speeds from rotor inboard tip vortices generating an upload on the horizontal stabilizer, in addition to tail buffet. The effect was mild and acceptable. The XV-15 control system made the aircraft much easier and pleasant to fly than the XV-3. The control runs for tiltrotors are, by their nature, longer than a helicopter's and more likely to suffer lost motion and excessive friction. The design and construction of the XV-15 sought to reduce such effects that had plagued the XV-3.

Top left: **This photo of the XV-15 reveals the downwash pattern under the tiltrotor with two fairly quiescent zones (the survivor in the water floating unmolested within one) with strong lateral flow beyond these, and powerful jets along the centerline projected off the nose and tail.** Jay Miller Collection

Bottom left: **The XV-15 was one of the most successful experimental aircraft ever developed and decisively demonstrated the practicality of the tiltrotor. Its lineage with the V-22 is readily evident.** Jay Miller Collection

Bottom right: **The XV-15 was the subject of military utility demonstrations, here during low-level nap-of-the-earth flight during the summer of 1982 at Fort Huachuca, Arizona. Adorned in combat colors, a chaff dispenser has been scabbed onto the aft portion of the sponson.** NASA via Marty Maisel

Right: **This time in a Navy finish, the XV-15 executes a landing in August 1982 aboard the Marine amphibious assault ship USS *Tripoli* (LPH-10) while underway. Nothing was found to be markedly difficult or different from normal helicopter shipboard operations.** NASA via Marty Maisel

The change in collective and cyclic function with nacelle angle was phased-in and phased-out mechanically and automatically during conversion and reconversion. The control surfaces were active in all phases of flight but naturally gained effectiveness as flight speed increased. During APLN turns a small amount of additional collective pitch was commanded automatically on the outside proprotor to combat adverse yaw, although pilot compensation with rudder was still required for properly coordinated turns. Like the XV-3, the new aircraft experienced chugging, although it was little more than an annoyance. Guest pilots commented that the XV-15 flew as well or better in VSTOL than comparable helicopters, and the same in APLN when comparable twin turboprop aircraft (particularly in SEO scenarios).

Conversion was a fairly simple procedure. During transition the pilot reduced or 'beeped' the rpm from 98-100% rpm in VSTOL to the 84% that was more quiet and the most efficient for APLN. The flaps and landing gear were raised during the acceleration. Above 160kts the nacelles had to be at 0° to prevent rotor overstress. During reconversion the beeping and flap extension process was reversed. Transition could be completed with moderate pilot workload in just 12 seconds. A unique landing approach could be flown using nacelle angle to control airspeed and power to control rate of descent. This maintained a level deck angle so that the landing point did not disappear under the nose as is common with a helicopter. A steep approach could be flown with the nacelles back at 95° for a high rate, nose low descent. A letdown at 1,000fpm and 50kts could be flown with ease. For STO, a 75° nacelle angle was used for optimum ground acceleration and climb out. At typical conditions the ground roll was just a few hundred feet.

Runstand ground operation with the first XV-15 commenced in January 1977. Hover trials began on 3 May 1977 with Ron Erhart and Dorman Cannon at the controls. The initial results were very positive. Perhaps recalling the XV-3 experience, the aircraft was moved to NASA Ames Research Center after just three hours of flight for full-scale wind tunnel evaluation. The second aircraft took up flight testing. The first nacelle forward rotation to 85° was performed on 5 May 1979. Full conversion with the nacelles at 0° followed on 24 July. This event was achieved after only 15 fight hours and with very little difficulty, Cannon and Erhart again flying, culminating many years of Bell tiltrotor research. The handling characteristics and ride quality of the XV-15 was very good. The conversion corridor proved generous, with an average width of 70kts, and did not required undue pilot attention.

Years of ground and flight testing followed by Bell, NASA, and the DoD. This explored all aspects of tiltrotor flight and produced an invaluable wealth of data. On 17 June 1980, the XV-15 set an unofficial speed record for rotorcraft of 301kts in level flight, becoming the first rotorwing aircraft without separate cruise propulsion to exceed 300kts in forward flight. An altitude of 26,000ft (7,925m) was eventually reached – another value normally unattainable by conventional rotorcraft. Few substantive changes were made to the aircraft. It was learned during flight test that the roll during sideward translation was undesirable for Nap-of-the-Earth (NOE) flight because of excessive bank angle. Consequently, a Lateral Translation Mode (LTM) feature was added to the controls that applied up to 4° symmetrical lateral cyclic on each proprotor to maintain wings level with the normal differential collective for sideward movement.

The XV-15 was landed at idle power, with a full reconversion. An autorotation landing sequence was also repeatedly demonstrated at altitude, with the engines at a sub-idle power setting. Actual autorotations to landing were

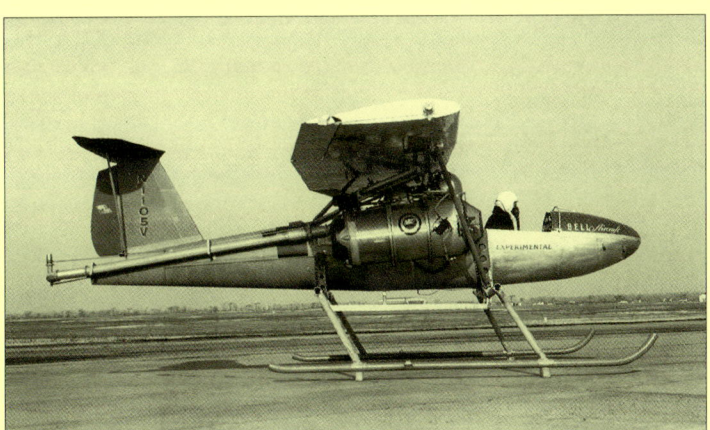

Bottom left: **The 2,056-lb Bell ATV first hovered free of a tether on 16 November 1954. The closest it came to a true reconversion was in forward flight with the engines rotated to vertical and then back to horizontal before a roll-on landing.** Jay Miller collection

Bottom right: **Although Bell's 3,100-lb X-14 was limited in flight envelope, it proved to be an invaluable test aircraft. For over 20 years it supporting data collection and pilot training for numerous VSTOL programs.** NASA Ames

Left: **The X-22 was a substantial departure for Bell at 16,755 lb maximum GW and four three-bladed ducted propellers. A series of drive shafts and transmission gearboxes, 11 in total, ensured that all propellers would turn if any the four engines failed.** Tony Landis collection

Bottom: **Vertol built and flew the first tiltwing aircraft with their Model 76 (VZ-2A). Much-modified during its ten-year career, the demonstrator eventual completed over 448 flights with more than 50 hours, including over 34 full conversions.** Jay Miller Collection

never performed, but the team was confident that it could be done safely. The demonstration saw a high 2,800-fpm (14-mps) descent, and the event was comparable to an autorotation landing in a heavy weight CH-53 with about the same level of pilot skill. It seemed clear that the maneuver would be the subject of simulator indoctrination in a fielded tiltrotor but would never be actually performed to a landing in training. This is not atypical of multiengine rotorcraft where autorotation landings are only conducted during certification trials. Line pilots seldom performed autorotations in anything but training helicopters.

The DoD performed XV-15 military utility demonstrations. More than 30 pilots, from both helicopter and fixed wing 'persuasions', flew the aircraft and their reactions were assessed. Even compared with a helicopter, the agility was most respectable. The ability to change speed rapidly (60 to 180kts in 10 sec), reposition quickly and precisely, turn very tightly in many scenarios, and control deck angle at will for nose pointing were all potentially telling combat attributes. The advantages of the tiltrotor design were found to offer enhanced survivability when compared with helicopters or fixed wing aircraft performing the demonstrated missions. Operations from a Marine assault ship particularly looked at deck edge effects. This was a concern with one rotor over the deck and the other partially or fully off the edge. The different downwash and ground effects was predicted by some to complicate aircraft handling in shipboard approach and hover. The XV-15 was launched and recovered repeatedly, including rolling takeoff and landings, with and without crosswinds. No difficulties were experienced and deck edge effect was easily manageable.

The aircraft also approached and hovered over a team simulating an external load hookup operation to evaluate handling, plus heating and ground wash effects on the personnel. The area directly under the aircraft was relatively calm due to the convergence of the rotor downwash plumes, facilitating a successful load hookup. A man walked around beneath the hovering aircraft with little problem in maintaining his footing. Downwash effect during a simulated sea rescue was also evaluated, again finding a quiescent zone directly under the aircraft. Experience showed that recirculation and groundwash did not lift ground material in a manner that reduced pilot visibility or otherwise impeded operations.

The design allowed very low-level NOE flight closely following terrain features for improved masking. Energy management was also of value during NOE where the nacelles were rotated forward when passing high ground to 'bunt' or push over. Airspeed increases and the nose is lowered, allowing closer conformity to the terrain for improved masking. The level aircraft deck with airspeed changes permitted less obstruction to forward vision and greater ability to ensure proprotor tip path clearance from obstacles. The rapid energy change with nacelle rotation also proved beneficial in such operations. The XV-15 demonstrated the ability to hover with a 30° nose-up attitude using a 60° nacelle angle, making unusually steep slope landings possible. The level deck angle possible by manipulating nacelle angle and longitudinal cyclic optimized scanning of a landing zone (LZ). This, combined with the ability to affect rapid energy change, could reduce exposure time during approach and departure by 40%. Simulated helicopter and fixed wing 'adversaries' found that the VSTOL advantages in air combat enhanced evasion potential.

Ship 2 was destroyed in an accident on 20 August 1992 after having logged 840.7 hours. A bolt had worked free in one proprotor control and allowed the collective pitch to runaway while in a hover. The machine immediately rolled inverted and impacted the ground with irreparable damage. Neither of the pilots was seriously injured. It was later found that the cotter pin retaining the nut on the bolt had not been installed.

The XV-15 was one of the most successful VSTOL experimental aircraft programs. It flew for more than two decades, executing more than 5,000 conversions with over 300 pilots. Demonstrations at the 1981 Paris Air Show, the Pentagon, and from the east parking lot of the Capitol Building in Washington, DC, astonished audiences. It made VSTOL look simple and effortless. No serious technical roadblocks had been encountered in the TRRA and the technology appeared suitable for a production application. The success generated considerable interest in the US armed forces and made the tiltrotor a natural selection to meet a military VSTOL requirement for a medium tactical transport. During the subsequent V-22 Osprey development program, the XV-15 provided a ready demonstrator and pilot training asset, plus a testbed for technology development and data collection. It was retired on 16 September 2003 with 679.8 flight hours.

Top: **Boeing Vertol did pioneering work with wings to off-loading the rotor during helicopter forward flight. Here a CH-46 has been fitted with such supplemental lifting surfaces, albeit small to keep download low.** Jay Miller Collection

Middle: **The extraordinary Model 347 gave Boeing an early look at large helicopter electronic flight controls and other design factors that would come into play on the V-22. The extensively modified Chinook included an advanced navigation system for instrument flight and vibration suppression features.** Jay Miller Collection

Bottom: **Boeing contributed to the 148,000 lb maximum gross weight XCH-62 program that yielded this massive three-engine, twin-tandem airframe before being canceled in July 1975. Although never to fly, the technology developed would carry forward to the V-22.** Author

Technology Base

Besides its tiltrotor demonstrators, Bell Aircraft was at the forefront during the heyday of VSTOL flight research in the United States. Vertol, and eventually Boeing Vertol, pursued advanced rotorcraft technology, including improved helicopter flight control, system integration, and simulation. One of their proprotor blade designs was tested briefly on the XV-15. These efforts contributed immeasurably to the V-22. The most visible products of this work were their test vehicles. These gave the firms considerable practical experience in solving the unique problems associated with such flight, and in testing these craft and their components. Design and manufacturing techniques, especially for gearboxes and composite materials, were other fallout. Less visible were the decades of analytical studies, ground tests and design proposals for such aircraft meeting a broad variety of military and commercial requirements. All this placed Bell and Boeing in a competitive position when a DoD tiltrotor project was launched.

The first vehicle to result from Bell's intensive VSTOL research was the self-funded Model 65 Air Test Vehicle (ATV) of the mid-1950s. The tiny demonstrator employed two turbojet engines installed on either side of the fuselage with mounts allowing them to be rotated through 90°. Although never to demonstrate full conversion, the ATV did provide instructive experience in recirculation and test techniques. Bell followed the ATV with the thrust vectoring X-14. Under NASA operation, the machine proved to be one of the most useful and long-lived experimental aircraft and trainer, allowing the industry to learn much about control and suck-down. Under the Army's high-speed helicopter research program in the mid-1960s Bell operated a modified YUH-1B with stub wings and two cruise turbojets. The quad ducted tilt-fan X-22A of the 1960s and '70s was developed with Defense funding. This unusual machine also lent considerably to research on VSTOL stability and control technology.

Vertol also contributed much to VSTOL research, building and flying the first successful tiltwing aircraft. This was the small VZ-2A, taking to the air in 1957. The Model 347 of the early 1970s sought improved handling qualities matched to greater maneuverability, and improved stability and control with a highly modified CH-47A Chinook. This incorporated stability augmentation plus basic attitude and heading hold features. A variable-incidence wing was installed atop the fuselage. The company had previously tested a CH-46 with small wings, but the 347 wing was large and included trailing edge control surfaces. The wing could rotate to

Top: **Especially significant in the evolution of the technology supporting the V-22 was the Boeing 360. The 30,500 lb helicopter produced a 20% improvement in payload, 30% increase in airspeed, and a 60% rise in productivity compared with similar rotorcraft.** Boeing via Jim Jagodzinski

Middle: **Composite helicopter construction was advanced through multiple efforts over many years, one being Bell's Model 292 Advanced Composite Airframe Program. It first flew on 30 August 1985 for a brief flight test program.** Jay Miller Collection

Bottom: **The LTV-Hiller-Ryan XC-142 tiltwing of the 1960s was the first tri-service VSTOL transport program in the US. It progressed well into development, with five vehicles built and extensively tested before ultimately being cancelled.** Jay Miller Collection

85° leading edge up for hover to reduce download. Control of rotor functions, wing incidence and the automatic flaps were integrated via the flight control system. A fly-by-wire (FBW) system was then installed, making the 347 the first helicopter ever to fly with full electronic control. The 347 work was a lead-in to the Heavy Lift Helicopter, the giant XCH-62A. Although this program was cancelled, it provided valuable design experience with advanced power transmission systems. Boeing went on in the 1980s to participate in the Army's Advanced Digital/Optical Control System program, producing the world's first digital FBW helicopter flight control system. The work was consummated with flight testing on a Boeing-modified UH-60.

Especially noteworthy was the Boeing Vertol Model 360 Advance Technology Helicopter of the late 1980s. Pursued as a company-funded technology demonstration, this was an entirely new design. Featured was extensive use of composites throughout in the low-drag airframe. The rotor blades had advanced transonic airfoils and the rotor incorporated elastomeric bearings for reduced vibration. The electronic cockpit design was dominated by six multi-function displays (MFD). Of greater note was the integrated dual automatic flight control system and flight management system. This included full-time three-axis stabilization with flight control law gains adjusted as appropriate for the flight condition to optimize controllability. Hover, heading, airspeed and bank angle hold options were incorporated along with coupled navigation modes.

Composite aircraft construction was a technology that was just maturing in 1981, but held promise of allowing a lightweight airframe but with optimized structural properties. In a few years Boeing Vertol would garner valuable experience with composites in with their Model 360 program that used the material in rotor and blades, transmission and control systems, and airframe. Likewise, Bell would increase its composite structures capability with the Model D-292 program funded by the US Army. This Advanced Composite Airframe Program sought to extend the technology for composite aircraft manufacturing with a dramatic reduction in weight, cost, radar signature, and survivability. The resulting light helicopter, with an airframe built almost entirely of these materials, first flew in August 1985. Besides the high strength-to-weight advantages, other benefits of composites were corrosion resistance – critical for a maritime aircraft – resistance to fatigue and crack propagation for long airframe life, and a reduction in assembly parts. Composites could also give improved ballistic and damage tolerance for reduced severity of damage from gunfire and shrapnel, and less vulnerability to foreign object and 'hangar rash' damage. In the event of a crash, the composite airframe failure modes tend to dissipate more energy than metallic structures, reducing the impact loads imparted to occupants.

A digital FBW control system, possibly using fiber optics to carry the signals, was planned for any new tiltrotor aircraft to eliminate the few vices observed in the XV-15. Flight control could be optimized to make the aircraft easier to fly and to tailor control for specific modes of flight. FBW also reduced weight by reducing or eliminating mechanical control runs. Digital avionics would increase reliability. Built-in test (BIT) features of avionics and aircraft systems would reduce maintenance time by allowing faults to be isolated and diagnosed quickly to speed repair, and reduce routine maintenance checks.

Throughout the 1970s and early 1980s Bell continued basic tiltrotor research and design studies, sometimes with government funding. Bell soon had a whole family of conceptual tiltrotor aircraft on the drawing board, among them medium transports filling a variety of military roles. Bell and Boeing amassed 9,000 hours of wind tunnel testing, with rotor designs and 27 scale aircraft designs models, and over 1,000 hours of flight simulation work.

Bell's conceptual design work included many that were modified XV-15s and others that were substantially larger but reflected the same layout. These began to incorporate more powerful engines, fuselage side sponsons for greater fuel capacity than the wing volume could allow, and a rear loading ramp. Engine infrared (IR) suppressors (IRS), aerial refueling (AR) probe, rescue hoist, ground mapping radar, and a gun also began to appear; desirable accouterments for tactical transports. External stores, including electronic warfare (EW) pods, rockets, and missiles, were seen in artist's concepts. Such stores were problematic because of the complex flowfields around the tiltrotor such that store jettison dynamics increased the potential of an impact with the airframe. Weapons would suffer from limited fields of fire at most practical mounting locations because of the proprotor arcs while in airplane mode. Wing mounts were considered impractical, but mounts on the lower quadrant of the fuselage, both forward and aft of the proprotor plane, were envisioned.

Bell's tiltrotor designs in response to DoD programs show the evolution that eventually produced the V-22. The D-314 series of designs, circa 1975, were essentially scaled-up XV-15s proposed under a US Army study called Spectrum. These were to fill a number of Army missions including medium lift transport. Another variant was much like the XV-15, although perhaps a bit wider, and intended to fill the Army's Special Electronic Mission Aircraft (SEMA) requirement. Artist concepts of this machine had a side-mounted side-looking airborne radar pod and under-wing EW pods. The D-323 was another XV-15 derivative, albeit with a V-tail and boxy fuselage, designed to fill a US Marine Corps air assault mission and featuring a rear loading ramp and pivoting wing stow feature. The D-316 and D-320 were intended for the USAF combat search and rescue (SAR) mission. The D-317 was aimed at the anti-submarine warfare (ASW) role and the D-318 the Navy Light Airborne Multi-Purpose System requirement. The D-319 series of aircraft were conceived in response to other Navy requirements. The GW of some of these proposed aircraft exceeded 18,000 lb (8,165kg) with cruise airspeeds of 250kts and ranges over 350nm, and could accommodate more than a dozen troops.

The best means of introducing such new technology as a production tiltrotor aircraft is via a military program. The government pays for development and testing, revealing and resolving fundamental risks and establishing a manufacturing base. By the 1980s the technological elements of a medium VSTOL transport was coming together. But, the DoD's work in the 1960s, that included the tiltwing XC-142 and Canadair CL-84 Dynavert, and the tilt-prop Curtiss-Wright X-19, had failed to yield acceptable designs and interest waned. The struggle for developmental funding during and after the Vietnam War, and a debate over aviation missions and aircraft types, was not an environment conducive to pursuing unusual designs that sacrificed performance for VSTOL. The services had to wait until political fortunes took another turn before again attempting to bring forth a VSTOL transport aircraft, with a multi-service approach considered the only practical avenue to such a venture.

A truly versatile and high performance, practical VSTOL tactical transport awaited the advent of high thrust-to-weight ratio engines with good fuel economy, advanced propeller aerodynamic and rotor system designs, electronic flight control with lightweight components, and lightweight composite materials. All of these technologies, building upon the previous decades of research, came together to make the V-22 Osprey tiltrotor aircraft a winning solution.

Top: **The Canadian CL-84 tiltwing was partially funded and evaluated by the US under a tri-service program. Although quite successful, the timing was just not right to carry the experimental design forward into a weapon system program.** Jay Miller Collection

Right: **Another X-plane that might have led to a tri-service VSTOL transport was the tilt-prop Curtiss-Wright X-19 that utilized the radial lift force produced by a propeller operating at positive angle of attack. Development was lengthy and the program ended abruptly with a crash during the first attempt at transition.** Tony Landis collection

EG
6418
13
EG
7668

U.S.AIR FO
8219
439 AW

XH-59A
21942
ARMY/NAVY/NASA USAF

Chapter Three

Birth and Hiatus

The JVX Program

In December 1981, Secretary of Defense Casper Weinberger announced DoD's intention to develop a multi-service aircraft to fill unspecified mission requirements for the United States Marine Corps, Navy, Army and USAF. If successful, it would be the first aircraft developed to serve in all four armed services. This was a time of unprecedented peacetime US defense budget growth and new weapons procurement, and the idea found ready consideration. However, previous attempts at developing multi-service aircraft, such as the F-111 and XC-142A, had proven unsuccessful or produced disappointing results. The requirements would have to be carefully tailored and receive the ready acceptance of all services.

A joint program would reduce costs by increasing commonality across all branches of the armed forces. A common airframe would help economize manufacturing through larger production lots for reduced unit cost. Joint training, logistics commonality, and possibly combined depot maintenance would cut operating expenses. A challenge in making any aircraft suitable for the Department of the Navy (DoN), Army, and Air Force were that their different maintenance concepts meant repair manuals, tools distribution and training had be somewhat different.

In their December 1982 report, the Joint Services Operational Requirements (JSOR) group identified ten missions that a multi-service aircraft could likely perform. An associated VSTOL study team, the Joint Technology Assessment Group (JTAG), included representatives of the services, NASA, and industry. It focused on vertical lift aircraft needs across the armed forces and examined all likely candidate concepts. They looked at advanced rotorcraft technology, high-speed helicopter designs, lift/cruise fan concepts, and tiltrotors. Evaluated was the Sikorsky XH-59A experimental helicopter with its Advancing Blade Concept coaxial rotor system and twin auxiliary engines. In its compound helicopter configuration the XH-59A reached 238kts in level flight and 25,500ft (7,620m) altitude. The technology was probably suitable only for a light attack machine or a utility aircraft.

The operational demonstrations of the XV-15 had proven more noteworthy. They appeared to illustrate that the technology was finally available to permit a practical military VSTOL transport. The tiltrotor held promise of doubling the speed and range over the comparable helicopters. Lower rotor loads and vibration levels promised higher mission availability with greater reliability, for lower maintenance and logistics demands than helicopters. This would mean fewer personnel required to maintain the aircraft, for additional savings. Although ostensibly more complex than a helicopter, the tiltrotor's operating costs were expected to fall between that of a fixed wing aircraft and a helicopter.

The results of the technology assessment recommended a tiltrotor as the most promising design for a medium-lift multi-mission transport aircraft. The formal Joint services advanced Vertical lift aircraft (eXperimental), or JVX, development program was born in December 1982 with operational requirements for the new system approved the same month. Based on Weinberger's initial announcement, a joint rotary-wing development program office had been established in December 1981 with the Army as the executive service. By this time the Marines had an increasingly critical need for an aircraft replacing the medium-lift CH-46E and the early model heavy-lift CH-53A and D aircraft. Their program, the V/HXM or just HXM (Helicopter eXperimental Marines), had been initiated in 1981. It soon became the cornerstone of JVX.

The V/HXM had sought an aircraft with an initial service introduction date of 1991, but the requirement for a new medium vertical-lift aircraft dated back to 1969 following the Medium Assault Study of the previous year. The program had undergone a convoluted history. An HX (Helicopter eXperimental) program in the

Facing page top: **The venerable CH-46 Sea Knight was a worthy mount in its heyday, but by the 1980s was becoming aged and expensive to operate – problems that were only exacerbated in later years. Its performance limitations were hampering the evolution of USMC tactics.** DoD

Facing page middle: **The MH-53 Pave Low provides the Air Force a vertical lift platform for special operation missions using a radar for low level terrain following flight. However, it commonly has to be partially disassembled and flown to the theater of operations onboard a C-5.** USAF

Facing page bottom: **A new rotorcraft technology examined for possible JVX application was the Advancing Blade Concept exemplified in Sikorsky S-69 (XH-59A) experimental compound helicopter. It was doubtful that the approach would be suitable for an aircraft the size of the JVX.** Author's collection

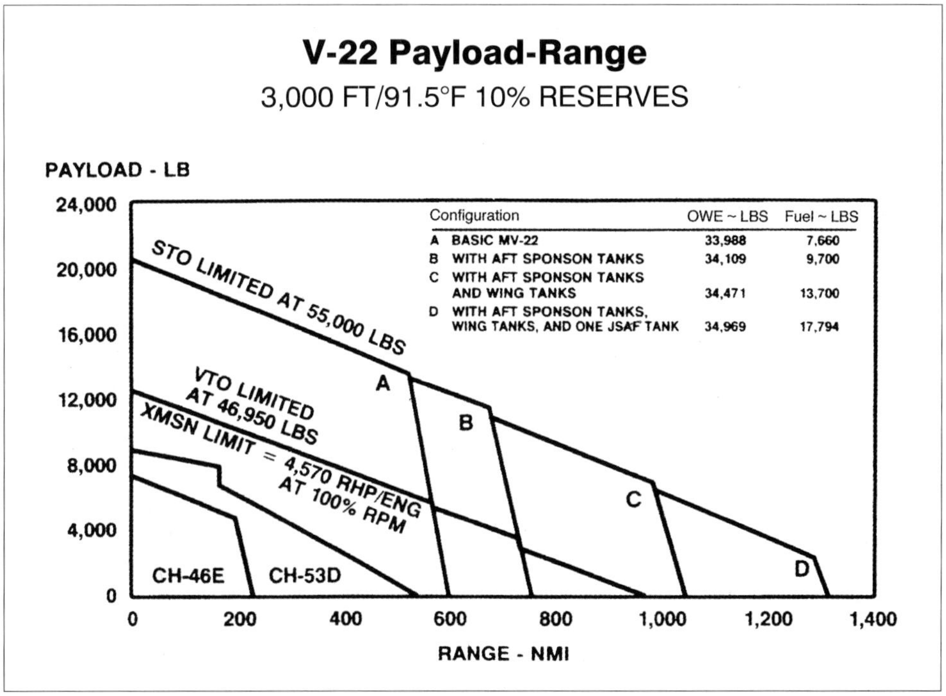

Right: **This idealized and now-dated diagram remains useful in revealing the range and payload advantages of the V-22 tiltrotor over the USMC helicopters it is intended to replace, entirely or in part.** Bell Helicopter

early 1970s had seen Bell propose the D-310 tiltrotor for the Navy and D-311 for the Marines. In 1972 the USMC had been directed to work with the Army on the UH-60 development, but their different requirements could not be accommodated in a single machine. The following year HXM was initiated and joined to the Navy's HSX (Helicopter Sea eXperimental) program, although the latter soon fell by the wayside. By the mid-1970s the Navy was seriously examining VSTOL to meet mission requirements. One would be a subsonic support aircraft, VSTOL A and suitable for the HXM, which emerged in late 1980 to which Bell proposed the D-321 and D-324. This, too, went away when funding failed to materialize. Late that decade the USMC again had to show that the UH-60 Black Hawk was unsuitable for their mission, justifying continuation of the HXM. Program initiation to develop the HXM was set for 1982, but was again delayed.

By the 1980s the CH-46 was badly dated and with no practical means of modification to meet the realities of the current combat environment. The last US model was produced in 1971. The E-model upgrade was begun in 1972 as an interim solution to the USMC medium-lift needs. In the 1980s the helicopters were undergoing a safety, reliability and maintainability improvement program to allow them to operate suitably until a replacement could be fielding in the 1990s. The CH-53D was not so bad off, but upgrades would not help to reduce the excessive cost of maintaining and operating the aging aircraft.

The HXM requirements translated to carrying 24 troops (vice 11 in the CH-46E), plus the two pilots and two crew chief/gunners. Its range was to be 200nm (vice the 46's 95 with a 30-min reserve), or a 100nm round trip from a ship, and to hover OGE at 3,000ft (914m) at 91.5°F (33°C). Consequently, the support ship could stand as much as 200nm offshore while the HXM quickly deployed troops inland. This would increase tactical surprise and help the amphibious ship to avoid enemy defenses. Alternatively, a cargo of 5,760 lb (2,613kg) was to be carried internally or 8,300 lb (3,765kg) externally. The aircraft would be faster and quieter, increasing the element of surprise, and

Top: **Another step in the evolution of the V-22 was the D-310, Bell's submittal for the Navy's HX program of the early 1970s. Like many of Bell's design concepts from this period, the 310 shared much in common with the XV-15 save for the 'butterfly' tail.** Jay Miller Collection

Middle: **The D-314E from 1975 was one of a series of tiltrotor designs intended to fill a number of Army missions. Although essentially a scaled-up XV-15, its V-22 lineage is evident.** Jay Miller Collection

Bottom: **One of the earliest Bell Boeing designs for the JVX had modest sponsons augmented with external drop tanks, a simple up-swept aft fuselage, a gun projecting from the nose, and a dog-leg AR probe.** Jay Miller Collection

would have the endurance to circumnavigate threats if necessary. The mandated 250-kts cruise airspeed would be a tremendous improvement over the Sea Knight's 105kts, not to mention the specified 3-hours endurance compared with the 46's 1.4 hours. An AR capability and two fuel cells with a total 15,950 lb (7,235kg) of fuel temporarily installed in the cabin would permit self-deployment to 2,100nm. The new aircraft was also to survive in a nuclear, biological and chemical (NBC) environment and withstand hits from 12.7-mm rounds. Two cabin-mounted and a forward-firing gun were called out.

The stated USMC requirements would mean a marked improvement to amphibious and land assault troop lift, resupply and casualty evacuation missions. The ambitious 250-kts requirement of the HXM virtually ensured that a conventional helicopter could not be found acceptable. Nearly all the requirements were more severe than those of other services.

Self-deployment to reach distant operating locales is generally not practical for helicopters. They are usually flown to their destination aboard Air Force C-5 cargo aircraft or via slow ship transport. Both require many hours to prepare the helicopter for shipment, with some partial disassembly, and ready it again for combat, impacting operational responsiveness. HXM self-deployment would reduce the DoD airlift burden and demands on airborne tankers.

The Marines also sought to greatly enhance their night assault capabilities. The new aircraft was to have a cockpit compatible with night vision goggles (NVG) and a Forward-Looking Infrared (FLIR) system to display surrounding terrain during night flight. Introducing updated navigation avionics would greatly improve crew situational-awareness and first-pass location of the LZ.

The JVX would perform shipboard operations. For a tiltrotor solution, the Navy *Tarawa*-class amphibious assault ship (LHA) set the maximum rotor diameter at 38.00ft (11.58m) for a lateral tandem rotor layout. This was a reduction from what some considered an optimal 43.0ft (13.1m) for a medium-lift tiltrotor. The requirement ensured 5-ft (1.5-m) from the deck edge for the aircraft's wheels, 12.7ft (3.9m) of rotor tip clearance from the island structure for the spot opposite the island, and 1.0ft (0.3m) APLN tip path clearance from the fuselage sized by missions requirements. The consequence was an increase in disk loading at typical anticipated gross weights to 17.5-23.2psf (85.4-113.3kg/m^2) for the V-22 compared with

Top: **The final external configuration of the JVX was shown in many Bell Boeing artist concepts such as this one emphasizing the Army medevac mission.** Author's collection

the XV-15's 13.2-15.3psf (64.5-74.7kg/m^2), impacting some aspects of performance. A means of folding the proprotor blades and stowing the wing to reduce aircraft dimensions was also essential. The folded dimensions were dictated by a maximum width to permit a sufficient number to be parked above and below deck, a maximum height of 19.0ft (5.8m) to clear the hangar door opening, and a vertical stabilizer ground clearance to avoid contact with shipboard items. Apart from the LHA, the JVX would have to operate from the *Wasp*-class LHD amphibious assault ship, the 'small deck' Air Capable Ships, the *Austin*-class LPD (Amphibious Transport Dock), and the *Whidbey Island*-class LSD (Dock Landing Ship) vessels.

The conceptual dimensions of the JVX were to allow 7 aircraft to operate 'spread' on the deck of the larger aviation amphibious ship (on the 6 helicopter launch spots and another on the bow). Some 17 others could parked to the side in the stowed configuration and 6 more below in the hangar deck for a total 30 aircraft aboard the ship. More fanciful layouts saw dozens aboard ship, but actual deployments would likely see only one JVX squadron (12 aircraft) on any vessel, and seldom all topside.

The Air Force special operations forces (SOF) mission would be in need of new aircraft by the early 1990s to replace remaining HH-3E Jolly Green Giant and the aging heavy-lift MH-53 helicopters. The Air Force mission, eventually under United States Special Operations Command's (SOCOM), included long-range, covert insertion and extraction of special forces and combat search and rescue (CSAR). The requirement for a new platform dated from the end of 1981, greatly influenced by the failed Iran hostage rescue mission using RH-53Ds. During the April 1980 mission the assault force had planned for two overnight hide and refueling-sites before the assault on Teheran because of the helicopters' range. The inability to directly insert sufficient forces into an urban environment at long range led to an overly complex plan that ultimately doomed the operation. With three times the range of the MH-53, the Osprey would need fewer aerial refuelings and help ensure mission covertness.

The USAF sought an aircraft that could have flown at comparatively high speed directly to Teheran from the carrier in the Arabian Sea in one period of darkness. The requirement was to carry 12 troops or 2,880 lb (1,306kg) of cargo to 700nm radius flying at 1,000ft (305m) and 250kts cruise and hover OGE at 5,000ft (1,524m) at 90°F (32°C) before returning at 500-1,000-ft (152-305-m) altitude with 10% reserve fuel. They also wanted to fly at up to 54,000 lb (24,494kg) using a STOL takeoff.

Bottom: **The Navy's SV-22 did not progress much beyond this painting. The sub-hunter would have seen a radar added in the nose and weapons on side pylons.** Author's collection

The MH-53Js Pave Low in what became Air Force Special Operations Command (AFSOC) were growing old, expensive to operate, and not responsive to mission needs in the coming century. The mission demanded more secure communications, countermeasures, day or night, all-weather, low-altitude navigation, and other special gear. However, the USAF variant was to be 90% common with the Marine aircraft. Special equipment was deemed essential to the special operations (SPECOPS) mission. Among the unique gear was radar for low-level terrain following/terrain-avoidance (TF/TA) flight at night and in instrument meteorological conditions (IMC), turreted nose gun and a ramp gun, rescue hoist, and specialized avionics such as a Downed Aircrew Locator System.

The Navy logistical support mission saw a need for the JVX beginning in 1991 to replace many aging helicopters. It could take over ship-to-ship transport, vertical on-board delivery, vertical resupply (VERTREP, delivery to vessels without helo decks), and over-the-horizon missile targeting. It was also seen as a potential carrier onboard delivery (COD) aircraft,

Top: **The full-scale fuselage mockup shows the canted forward bulkhead that helps prevent a flipover in the event of a forced landing with substantial forward velocity. Note also the FLIR mount projecting from the bulkhead.** Jay Miller Collection

Left: **Mockups like the empennage allowed trial fit of items like the hydraulic and electrical lines seen here. The need for mockups has later virtually eliminated by computer-based design tools.** Jay Miller Collection

Facing page top: **The cockpit of the FSD Osprey's generally included test equipment like the display above the center of the glareshield and the control panel at the base of the short center console. Note the tall vertical TCL to the left of the starboard seat.** Bell Helicopter

Facing page bottom left: **The TCL developed for FSD, detailed here, came in for some criticism. Most significant was the upright grip with short movement arc. Compare with the final design shown on page 93 in this volume.** Author's collection

Facing page bottom right: **This artist impression of the JVX cockpit early in the program was very much like that finally realized. Most significant is the TCL design seen here that was to be fundamentally changed. Much later the design would revert to that shown.** Author's collection

LEFT SEAT
MAX 40 POUNDS

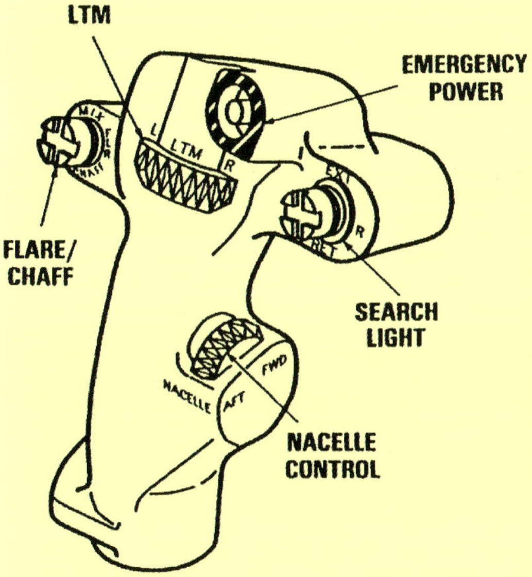
LTM
EMERGENCY POWER
FLARE/ CHAFF
SEARCH LIGHT
NACELLE CONTROL
NACELLE
AFT
FWD

replacing the venerable C-2 Greyhound, but providing the same cargo capacity to surface combatants equipped only with a helo deck. The JVX would likely not have the 'legs' of the C-2, but would probably carry as much and not require catapult and arresting gear. Cargo is commonly delivered to a carrier via the C-2 and then distributed to other ships in a battle group via helicopter VERTREP. The JVX could eliminate the intermediate step. The mission that got the most JVX attention was CSAR, then served by the HH-3 Sea King that also supported SPECOPS. The HH-3 was not well suited to these missions and the Navy had established the Combat Aircrew Rescue Aircraft program to find an alternative. The aircraft would require a crew of five flying to a range of 460nm with four rescuees and a cruise speed of 250kts. It had to hover OGE in hot/high conditions for 15 minutes at the mission midpoint. The JVX would allow a SAR aircraft to go farther to reach survivors, search a larger area longer, or pick up a greater number of rescuees. These and other missions were part of the Variable Mission Aviation Platform program for which the JVX was considered.

In March 1985 the Navy issued a preliminary requirement for a VTOL anti-submarine warfare platform. They announced in May that a derivative of the JVX would be replacing the S-3B Viking carrier-based ASW airplane and the SH-2 Seasprite helicopter. The JVX could operate from the helicopter decks of other surface combatants, freeing up aircraft carrier decks for more strike aircraft. The speed, range and endurance of the tiltrotor, combined with its greater payload gave it many advantages for the mission. The aircraft could get to the patrol area faster than a helicopter, search an area more quickly, and stay on station longer. Search patterns could be much tighter than a fixed-wing patrol aircraft. Dipping sonar would be deployed or passive sensors dropped. The JVX could set down on arctic ice while its dipping sonar hung down through holes in the sheet. This would allow prolonged missions hunting subs under the ice yet saving fuel.

Minimal changes were to be made to the baseline JVX to accommodate ASW. Conceptual designs showed the belly hatches normally used for external load hooks turned into a sonobouy dispenser forward and dipping sonar aperture aft. Up to 60 sonobouys might be carried and deployed from a hover or at very low forward velocity. Consequently, the units did not need to be as rugged as those deployed by the S-3 or P-3 Orion, and could be recovered and reused for cost savings. This 'soft-deployment' reduced sonobouy cost, but the ability to recover the unit also meant that they could be made more powerful and sensitive. Initial concepts suggested an UYS-2 acoustic processor and electronic support measures suite. For surface engagements an

Top: **Many JVX wind tunnel models of various scales and complexity were tested at numerous facilities. Here a full-span, 15% scale powered model is run in the Boeing-Vertol tunnel.** Author's collection

Bottom: **A full-scale V-22 proprotor was never tested in a wind tunnel or on a whirl stand prior to turning on the Ground Test Article. This image is of a 2/3rd-scale (25-ft diameter) version in the NASA Ames 40x80-ft tunnel with mocked-up wing section beneath to measure download.** NASA

Top: **The Ground Test Article allowed Bell to test its wing, engine, and proprotor design before a Boeing fuselage was available. The raised runstand permitted nacelle rotation down to 0° during ground runs, some that continued for many hours.** Jay Miller Collection

Bottom: **The assembly hall in Ridley Park, Pennsylvania, in August 1988. The fuselage of ship 2 was about 90% complete and 3 (foreground) was still waiting for its empennage.** Author's collection

APS-137 inverse synthetic aperture radar was added in an extended nose fairing. Weapons racks on the side sponsons were to carry Harpoon or Penguin anti-ship missiles, two to four Mk 46 or 50 torpedoes, or mines. An artist concept also showed torpedoes mounted under the inboard portions of the wing on both sides of the fuselage. Several AIM-9 Sidewinder air-to-air missiles could possibly be mounted on the forward fuselage for self-defense. And, external fuel tanks of up to 600 US gallons (2,270 liters) were considered. Protruding windows for personnel to visually scan the ocean surface were to be added in the aft fuselage. Although the ASW variant did not become part of the initial JVX requirements, it remained additional justification for developing the aircraft.

The armament considered for the ASW variant were not the limit of the studies of JVX weaponry. Various weapons and mounting schemes were considered for the turreted gun in the nose that some of the services wanted. Two Stinger missile tubes, for air-to-air engagements, mounted on both sides of the forward fuselage were also conceptualized. This work did not progress far as weight reduction concerns and the need to get the basic aircraft flying took precedence.

The US Army had various corps-level long-range combat logistics, utility, and air assault missions, and SEMA, for which the JVX could be suitable. It would almost certainly also find use with Army Special Forces. The SEMA requirement mandated good performance up to as high as 30,000ft (9,144m). Potential growth missions that were suggested were an airborne command and control center, and a communications relay platform. Some of these requirements had short-term need dates. The Army later stated a need for 24 of the machines to perform medical evacuation (medevac) duties. Basic requirements were for a load of 10,000 lb (4,536kg) of cargo or 24 troop, or 12-18 litter-borne casualties. The speed and range of the JVX would allow wounded to be carried directly from the forward battle area to rear area hospitals in 16 minutes instead of to the aid station just behind the front. By helicopter, it would be 14 minutes to the aid station and 27 minutes to the hospital. The aircraft was to be held at a 4,000-ft (1,219-m) hover in 95°F (35°C) temperatures, fly 250-kts cruise with a 300nm radius, and possess a 2,100-Nm ferry range. It was to hover OGE with either 8,300 lb (3,765kg) of internal load or 10,000 lb (4,536kg) on two external hooks. The new aircraft was expected to displace such platforms as the OV-1D, RV-1D, and a few specialized electronic warfare helicopters, beginning in 1993.

Even before the JVX had flown the manufacturers produced concept art of the aircraft in United States Coast Guard colors. Although

Left and below: **The fuselage of aircraft 1, the first airworthy V-22, arrived in Texas on 26 January 1988. Its vertical tails had been removed for clearance. The fitting at the top of the fuselage, on the stow ring, is to allow a crane to lift the fuselage onto a flatbed truck.** Both Jay Miller

the USCG had not openly expressed any desire for the aircraft, the range and speed of the JVX would clearly be a boon to their SAR mission.

On 15 January 1985 the JVX was designated the V-22 Osprey. The name invoked images of the aquatic bird of the same name that could hover over a point on the ocean and then swoop down onto its prey. The Marine version of the aircraft was designated the MV-22 and the USAF variant the CV-22. This was the opposite of expectations in that Marine helicopters typically used the CH prefix (cargo helicopter) and the USAF special operations helos MH (multi-mission helicopter). However, aircraft carriers were designated CVs, and having another weapon system in the DoN with the same designation was unacceptable. The USMC Osprey was to be the MV-22A (M for Marine), leaving the USAF with CV-22A, the C apparently consistent with the Cargo designation of fixed-wing transports. The insistence that a V appear in the dual designation left the Navy with HV-22, or Search and Rescue Vertical. The ASW version was dubbed the SV-22A. The Army had initially planned to buy the MV-22, apparently unconcerned with the alphabet soup.

The USMC stated a requirement for 552 MV-22s. The USAF saw a need for 80 CV-22s replacing the MH-53J Pave Low and some MH-130E Combat Talon II and MC-130P Combat Shadow transports and tankers (the Osprey requiring less tanker support) plus some MH-60Gs. The Navy perceived a need for some 50 HV-22. Perhaps as many as 231-365 Army MVs could be needed. Beyond that there was speculation that as many as 350 CV-22s could find roles in the Air Force and up to 300 HV-22s in the USN. Later estimates suggested that between 433 and 630 aircraft could potentially be sold abroad. At various times the possible number to be manufactured varied between 657 and 1,198. The development funding split between the services was to be 50% paid by the DoN (including the USMC), 34% by the Army and 16% by the USAF.

Schedules combining the various need dates of the services pointed to initial deliveries in 1987 and USMC initial operational capability (IOC) of 100 aircraft in 1991. This would clearly require an accelerated development program, and this was justified on the basis of the XV-15 having already suitably demonstrated the viability of tiltrotor technology. This and the other likely salient features of the aircraft, such as composite construction, a 'glass' cockpit, and digital fly-by-wire flight controls, were weighed. The incorporation of such advanced technology was to enhanced crashworthiness and survivability, provide greater multi-mission effectiveness, reduced pilot workload, and permit easier insertion of future system upgrades. Because all these features were already well along in development or in other aircraft, the technological risk was considered low. Based on that assessment a prototype of the proposed aircraft, a 'YV-22', was judged unnecessary.

Bell was already preparing a proposed 35,000-lb (15,876-kg) D-315 tiltrotor for the V/HXM competition. The design held promise for carrying 23 personnel at 15,000-25,000ft (4,572-7,620m) altitude and 266kts maximum airspeed to 200nm radii for the ship-to-shore mission. With a 5,000 lb (2,268kg) payload the range was to be 1,500nm employing VTOL, or 2,400nm with STOL. The 2,030-shp (1,515kW) General Electric T700-GE-401 was the preferred power plant. While the civil transports Bell was conceiving at the time had two of these engines in tip nacelles and a third in the fuselage, the military transport design had four engines with two in each nacelle. The fuel cells were placed in the wing and the landing gear in side sponsons. This aircraft would have had a V-tail and a straight wing that could be pivoted adjacent to the fuselage for below-deck stowage. Originally the V-tail, latter replaced with a T-tail, would have folded down for hangar deck clearance.

A more refined Bell V/HXM design, designated the D-327, emerged in 1981. This had the T-tail, a boxy fuselage with aft loading ramp, seating for 24 troops, and a pivoting wing. The wing featured a slight forward sweep. Only a single General Electric T64-GE-416 3,700-shp (2,760kW) turboshaft engine was installed in each wingtip nacelle. A 4,821-shp (3595-kW) version of the engine, the T64-GE-T5E then in development, was also considered as a growth possibility. The aircraft structure was to use composites extensively.

The V/HXM request for proposals (RFP) was suspended in favor of the JVX. The JVX missions, based on the JSOR report, combined the USMC, USN, USAF and Army needs. The demanding requirements could clearly not be performed by a conventional rotorcraft. The predominate features were:

Performance

- 250kts continuous cruise airspeed with a dash of 275-300kts below 500ft (152m) altitude
- -1 to +4G normal load factor for maneuvering and threat evasion (+3G VTOL)
- 200-300nm combat range (700nm USAF), 1,400nm tactical range, 2,100nm self-deployment (unrefueled)
- hover at 3,000ft (914m) OGE in 91.5°F (33°C) temperatures (4,000ft and 95°F/35°C USAF) with 8,300 lb (3,765kg) external load
- flight to 26,000ft (7,925m) with a 40,000ft (12,192m) ceiling
- fly away from an OGE hover under any condition after losing an engine

- maintain SEO altitudes of at least 15,000ft (4,572m) at full payload and fly at cruise speed to a landing
- power-off glide or autorotation to survivable emergency landing in the event of total power loss

Accommodations
- minimum crew of three
- seating for 24 troops with full combat kit or 12 litters
- 20,000 lb (9,072kg) roll-on/roll-off cargo for short-haul missions
- 880 lb (399kg) payload with 460nm radius
- 10,000 lb (4,536kg) sling-load to 50nm on a single belly hook, 15,000 lb (6,804kg) on two hooks
- airdrop of cargo via parachutes

Weights
- 31,768 lb (14,405kg) empty (guaranteed)
- 47,500 lb (21,546kg) maximum VTOL
- 55,000 lb (24,948kg) maximum STOL
- 60,400 lb (27,397kg) self-deployment

Operational Considerations
- AR capability
- armament of 2-4 air-to-air missiles and two 20-mm or 40-mm cannons
- countermeasures for enhanced survivability
- 'fold' to reducing dimensions for movement aboard a amphibious ship
- 'unfold' and made ready in 45-kts wind on 12° slope
- operate from amphibious ships and austere short strips ashore
- high reliability, availability and maintainability (RAM), 5.52 man-hours/flight-hour (guaranteed, 2.62 goal)
- operate in climates ranging from arctic to tropical
- operation in sand, dust, snow, moderate icing (45 minutes) and salt spray environments
- operation under NBC warfare conditions

The demanding RAM requirement, if achieved, would allowing fewer aircraft to fill mission needs and with less manpower to keep them flying. It was expected that the JVX would be 30% more reliable than older military aircraft and 100% that of helicopters. The same number of personnel manning a 12-aircraft CH-46 unit could maintain a 15-aircraft JVX squadron. This and a smaller number of aircrew were to save on operating cost. The JVX capabilities would permit a Marine landing with 30% fewer aircraft operating from the 'big deck' Aviation Ships.

Top: **Ship 1 performs an engine run on the test stand while the GTA waits on another partial runstand beyond. The aircraft was raised on jacks beneath the tires and then lowered onto the stanchions under the wings.** Bell Helicopter

The D-327 became the starting point for Bell's JVX proposal. With the T64-GE-T5E, performance would be boosted to better fit the USAF long-range mission. Other features incorporated to meet that role included a FLIR and radar installed in the nose, an AR probe, and a 7.62mm gun. Concept drawings also showed a cluster of four Stinger missile launch tubes on the forward fuselage. The D-237 was also suggested as a COD. A variant with an H-tail appears to have been the last step before the JVX proposal. Although the XV-15's H-tail enhanced directional stability and reduced weight, its carry-over to the JVX was advantageous in reducing aircraft vertical height for hangar deck stowage without a complex and heavy tail-fold feature.

The JVX request for proposals was issued in December 1982 for a 15.5-month preliminary design effort. The intent was to select two contractors to compete in a 23-month preliminary design portion. A 'simulator fly-off' would follow with the modeled designs evaluated against JVX missions. The winner would then be chosen. In the summer of 1982, plans envisioned fabrication of the first aircraft beginning in

Right: **An interesting perspective of ship 1 on the Arlington runstand showing the no-skid strip pattern.** Jay Miller Collection

Both pages: **Rollout of the first V-22 Osprey was a red-letter day in rotorcraft history and the struggle to introduce VSTOL aircraft into service. The 23 May 1988 event was staged at Bell's Arlington plant using ship 1 temporarily adorned with water-soluble combat colors.** Bell Helicopter (this page, left); others Jay Miller

December 1986 and first flight the following January (soon altered to August 1987). Initial deliveries were to be in July 1991. The entire development program was expected to cost $2.3 billion. Bell Helicopter, Boeing Vertol, Grumman, Lockheed, Aérospatiale, and Westland all expressed interest in the competition.

In line with new acquisition policy, the DoD encouraged Bell Helicopter Textron, with its tiltrotor background clearly a leader in the competition, to find a partner in preparing its proposal and executing any follow-on development. The two companies would then compete separately for annual production lots. The Pentagon felt that teaming would benefit the program by drawing upon a broader technology and experience base while reducing the financial risk for any single company. Teaming would bring Bell's tiltrotor expertise together with another firm's background in large aircraft possessing complex systems and avionics, helping to guarantee a better product. Consequently, Bell and Boeing Vertol (the latter eventually becoming Boeing Helicopter Company and later still the Boeing Defense & Space Group, Helicopter Division) formulated a teaming arrangement in the spring of 1982. The Bell Boeing Joint Venture or 'Tiltrotor Team' quickly began work on their JVX proposal.

The teaming had its down side. These were the only two US defense contractors with a large body of experience in VSTOL rotorcraft concepts. Theirs would likely be the only team to submit a competitive proposal. If left to develop proposals separately, additional innovation may have resulted. Tensions and duplication were engendered by the intent to have the companies compete for the production contracts. Each would assemble jigs and tooling to produce the entire aircraft by itself, with an attendant increase in costs. One of the firms would be guaranteed 40% of an annual production share, but both would compete for the more lucrative 60%. This was intended to create quality and price competition while ensuring production economy and suitable profit for each. Having dual sources for critical weapon systems was then a DoD objective.

The Bell Boeing team conceived the Model 901-X, what was essentially a scaled-up XV-15 based on the D-327. The 6-ft (1.8-m) high, 858-ft^3 (24-m^3) cabin and shipboard requirements sized the aircraft. Studies of engine arrangements looked at 11 variations of tilting engines and engines fixed to the wing or fuselage. The tilting engine was favored because it yielded 1,971 lb (894kg) more payload. Like a helicopter, the aircraft commander was to occupy the right seat because this is the required position for shipboard landings with the aircraft approaching the ship from astern and off the port side, opposite the ship's island. This provides the best view of the island, aircraft and deck equipment during the approach. The proposal included a helmet-mounted sight and hands-on-stick function control as part of the integrated avionics.

The design had a control surface layout identical to that of the XV-15, with segmented flaperons, twin rudders and an elevator. Unlike the XV-15, the flaperon segments on one side were to always move together for both flap and aileron functions. The flight control scheme was also identical to the XV-15. But, by adopting a FBW design, the XV-15's mechanical conversion/reconversion control system was eliminated. Tailoring of the flying qualities and handling qualities could be more easily accomplished with response feedback, pilot input 'shaping', and automatic control features. This optimization should be able to overcome some of the less desirable XV-15 characteristics such as sluggish vertical and roll response in hover, lateral darting during IGE hover, and large torque transients in airplane mode maneuvering. The XV-15's 'chugging' was also addressed with electronic controls using a rotor governor feed forward signal. Poor ground handling of the XV-15 was resolved with nose wheel steering in the V-22.

The V-22's engines were to be provided as government furnished equipment (GFE). The DoN, as the largest prospective purchaser, led the selection and negotiations for the power plant. A turboshaft engine in the 5,700shp (5,250kW) class appeared to be required for a twin-engine JVX to meet the SEMA and SOF missions. However, available engines met only the less-demanding Marine requirements. Fortunately, such a power plant was the hopeful outcome of the Army's Modern Technology Demonstrator Engine (MTDE, later just MTE) program. The objectives included 5,000shp (3,730kW) with 28-30% improvement in specific fuel consumption (sfc) to 0.43 lb/hr/shp (0.14kg/hr/kW) at a sea level cruise power setting of 3,000shp (2,240kW). To achieve this performance, high inlet turbine temperatures and pressure ratios were set as goals. Reduc-

tion in weight, increase in reliability, and improved maintainability were also sought. Each US turbine engine manufacturer expressed an interest in MTE. By early 1984 the Army's requirement had evolved to 5,600-shp (4,175-kW). Should the program yield suitable technology, the three primary services were expected to jointly fund development for initial production in 1986.

By the time the Bell Boeing team was preparing its proposal it appeared that MTE would not yield a suitable engine until the early 1990s. Consequently, an uprated version of the 4,855-shp (3,620-kW) General Electric T64, called the T64-GE-717, was considered as an interim power plant in the first 60 aircraft. This option might provide the power but not the sfc, reliability, and other MTE specifications, and also entailed additional costs for a future modification to the final engine design. Likewise, a version of the Avco Lycoming T55 was under consideration. However, it later developed that three engine manufacturers – General Electric, Pratt & Whitney, and Allison – assured the airframe manufacturers that prototypes of their MTE power plants would soon be available. Consequently, the T64 option was abandoned in the first half of 1985 and the first flight of the JVX slipped six months to February 1988 to accommodate the promised upgraded engines. Both the P&W PW3005 and General Electric GE 27, each derived from MTE, were being considered. As the JVX conceptual design was rapidly congealing, the choice of engine remained open.

The Bell Boeing proposal submitted on 17 February 1983 was the only response to the RFP. Seeing that a tiltrotor was favored and that the Bell Boeing team was the clear leader, the competition chose not to participate. This outcome initially prompted the DoD to consider canceling JVX, but they decided to proceed with a single design. The evaluators favored the proposal and the team was awarded a $68.7 million Stage 1 preliminary design contract on 26 April 1983. The funding was split equally between Bell and Boeing.

The contractors immediately launched into extensive analysis, trade studies, and ground tests. This included extensive wind tunnel work and scaled proprotor system whirl stand tests with early measurements of wing downloads. Over 4,700 hours of wind tunnel tests were performed during this period. The confidence gained from the XV-15 led the team to forgo full-scale proprotor tunnel tests.

To meet the SEMA mission and assist in matching the USAF range requirement, a JVX B configuration was conceptualized. This differed from the A configuration in possessing cockpit pressurization and larger proprotors for improved high altitude performance. When the Army withdrew from the program the Navy ordered termination of work on JVX B in September 1983. Proceeding with a single USMC configuration that could still meet USAF needs reduced the design workload and eased a looming weight problem. However, the USAF

This cut-away drawing permits identification of some V-22 features. Bell Helicopter

and USN models were to have uprated transmissions and higher proprotor tip speeds.

Stage 1 was followed in May 1984 with award of the Stage 2 preliminary design contract. This covered trade studies, initial development work, and risk-reduction design refinements. Piloted simulations ran to 500 hours. Further wind tunnel testing brought the total hours to over 7,000 in nine facilities with nine models. Extensive lab testing of the graphite/epoxy composite material provided data for detailed design. This ranged from small coupon tests to trials with large assemblies. A full-scale wing torque box structure was constructed and subjected to ground testing to substantiate this design. A wing/body segment with landing gear backup structure was built for ground loads trials, followed by fuselage forward and center sections for further static testing, and windshield bird strike tests. Detailed design of long-lead components was undertaken to reduce program schedule risk. The entire preliminary design testing effort was termed DT-I (development test).

The team submitted their Full-Scale Development (FSD) proposal in August 1984. Detailed discussions in September saw changes and another submission in February 1985. Among the changes was deletion of the nose gun until after production had commenced in the interest of keeping program costs within limits. The government also reduced the test aircraft from seven to six, plus two non-flying static test articles (STA) for structural loads and fatigue testing and a ground test article (GTA). As a consequence of these and other negotiations, pilot production was reduced to 12 from 18, while annual Lot 1 and Lot 2 (originally 42 aircraft) were plussed-up. The first production machine was still to be delivered in 1991. Plans at that time were for ten annual production lots through 2001 for 913 machines: 552 for the USMC, 231 for the Army, 80 for the USAF, and 50 for the Navy. This was expected to see a peak of 132 aircraft produced per year. This may have been the last aircraft program anywhere to consider such high numbers that were beginning to appear increasingly fanciful.

Go-ahead for FSD was expected in July 1985, but was slipped into October. This was probably to allow completion of the Cost and Operational Effectiveness Analysis (COEA) submitted that month by the contractor team and the USMC. The Navy then requested revised proposals from the prospective engine manufacturers and this delayed the process further. Final engine selection had still not been made, and it came down to a choice between the General Electric GE 27 demonstrator engine, the Pratt & Whitney PW3005 development of the PW3000, and the Allison experimental 501-M80C. Finally, in December 1985, Allison's entry was selected. The choice proved controversial.

The Allison engine, soon renamed the YT406-AD-400, was derived from the 8,079-shp (6,025-kW) T701 developed for the Boeing Vertol XCH-62. This was, in turn, a modified version of the 5,975-shp (4,455-kW) T56-A-427 then under development for an upgraded E-2C Hawkeye. The baseline 4,000-shp (2,980-kW) T56 power plant had been in service since the late 1950s and had accumulated over 130 million flight hours to that time on such aircraft as the C-130, P-3, E-2C and C-2A. Hence, the T406 offered commonality with other military turboshafts and unquestioned core components dependability. The Allison engine had an expected standard day maximum horsepower rating of 6,150-shp (4,585-kW). The attraction of a potentially higher-than-necessary power rating balanced greater weight than the competitors and slightly lower sfc. The fuel economy deficiency meant that the USAF range requirement would go from 700nm radius to 550.

Because of the anticipated large buy of engines, the Pentagon directed Allison to select a company to bring along as a second source to ensure deliveries and beneficial competition. In March 1986 Allison chose P&W to work beside them during development and prepare to manufacture the engine. Both would compete for production contracts.

Top: **The first V-22 taxies out at the beginning of its first flight from Arlington Municipal Airport on March 19, 1989.** Jay Miller

Middle and left: **First flight of the V-22 was at the end of a protracted and frustrating preparation period that placed the program behind schedule. Last-minute changes are evident by temporary data acquisition system cables and lines taped to the forward fuselage and nacelles.** Both Jay Miller

Right: **Test pilots for the maiden flight of the Osprey were Bell's Dorman Cannon and Boeing's Dick Balzer. Both were very experienced, with Cannon having contributed to the XV-15's success. Shortly after this photo was taken, Balzer died as a result of a freak hunting accident.** Author's collection

Bottom: **All but two of the first dozen V-22s were assembled at Bell's Plant 6. Aircraft 1 and 2 are nearing completion in this image.** Jay Miller Collection

The final FSD proposal was approved and Bell Boeing was selected in April 1986 to proceed with development of the V-22. The full seven-year, $1.714 billion, fixed-price contracts were signed on 2 May. Consistent with new DoD procurement direction, the documents stated that if costs exceeded $1.534 billion the government would assume 40% of the overrun, not to exceed $1.714 billion (soon raised to $1.81 billion). Beyond that the contractors would assume all the burden of any overrun. The idea was to generate cost-control discipline and to shift more of the program risk from the government. With the addition of associated contracts for the engines and other GFE, the overall V-22 program rose to more then $2.5 billion. The unusual fixed-price contract for development and the first three low-rate production lots compelled the contractor to make changes to the proposed design to reduce risk, especially with respect to potential weight increases. To meet the guaranteed empty weight a crushable floor, intended to reduce crash impact forces imparted to occupants, was eliminated.

Using their own funds, the contractors had been hard at work since June 1985 with continued research and development (R&D) aimed at further risk reduction. The work split was decided to be Boeing responsible for fuselage/cockpit, empennage, overwing fairing, landing gear, electrical, avionics and hydraulic systems, flight controls, and aerodynamic performance. Bell had the wing, rotors, engine integration and drive system, conversion system, and air vehicle dynamics. Boeing was to handle the static loads and drop article STA while Bell was responsible for the fatigue article STA and the GTA.

The DoD gave formal go-ahead for FSD in December 1986. At the time, the first flight was still planned for February 1988. Delivery of the pilot production aircraft was to begin in April 1992 and running over 13 months. Lot 1 was again changed, this time to 24 aircraft. A unit cost of $16 million per aircraft was estimated at the time, probably based on a wholly optimistic purchase of 1,088 machines. The Army was having difficulty fitting the V-22 into its budget and the USAF had already reduced its buy to 55 machines to replace 89 helicopters and C-130s. Loss of these orders alone would increase the unit cost to $20 million. If one included the cost of the government-furnished articles such as the engines, $30 million seemed a more reasonable figure. At time of rollout of the first aircraft in May 1988 the DoD was quoting $34.5 million. A full buy of over 900 aircraft could see the program grow to $25-30 billion. All these numbers became a jumble of claims and counterclaims between the contractors and various Federal agencies when the program ran into trouble.

Because the DoN would be procuring the lion's share of Ospreys, the Navy had been selected as the new lead development service in January 1983. The Navy's aircraft development arm, Naval Air Systems Command (NAVAIR), then became Bell Boeing's customer. An additional $4 million was found in the Navy budget to study the SV-22 ASW derivative. At that time it was anticipated that full-scale development of this variant would begin in late 1988 with a 1995 first delivery date. The Army withdrew from the program entirely in 1988 because its budget had to concentrate on more pressing aviation programs.

After decades of work, the technological and political 'stars' had finally aligned for an American VSTOL transport aircraft to enter development with a high potential for deployment as a military weapon system. Bell's long championing of the tiltrotor, with Boeing not far behind, and government sponsorship of R&D was to finally pay-off with the first production tiltrotor apparently only a few years away.

V-22 Full Scale Development

The V-22 Osprey rapidly shaped up as one of the most advanced aircraft ever attempted to that time. It used a triple-redundant digital FBW flight control system for stability augmentation and enhanced handling qualities, and also to eliminate mechanical control complexity through the rotating wing stow interface. This also permitted a Conversion Protection System to be readily incorporated to automatically ensure the pilot remained within the corridor. A Full-Authority Digital Electronic Control (FADEC) system gave totally electronic control of the engines and optimized performance. The use of FBW and FADECs greatly reduced aircraft weight by eliminating mechanical control runs. With digital flight control still relatively new, each FSD aircraft featured an Analog Backup Computer for flight controls and an Analog Backup Engine Control to be used only in flight test.

The 'muscles' behind the flight controls was a triple-redundant 5,000-psi (345-bar) hydraulic system. The V-22 was the first aircraft to use such a high-pressure system. It was adopted to allow smaller pump and actuator dimensions, and smaller diameter tubes, for reduced weight. However, it also compelled use of costly titanium and steel hydraulic lines.

The 'nervous system' of the aircraft included dual-redundant data buses and mission computers. These integrated the avionics suite made up of 55 'black boxes' and 59 subsystem assemblies. A 'glass' cockpit was dominated by MFDs with many display formats for tremendous information availability.

A center stick/cyclic was retained, vice the sidestick controller evaluated briefly on the XV-15 when a military tiltrotor was initially under consideration. It addressed the issues of center stick motion limited by the pilot's legs, seat or control panel, or obscuring panel displays. The sidestick was found suitable and even superior to the conventional cyclic in some situations. However, the center stick was felt essential in VSTOL to recognize the cyclic position precisely and judge remaining control margin for operations like taking off from a slope after a large change in center of gravity (for example, following unloading of troops and cargo). A major departure of the Bell Boeing design from the XV-15 was deletion of the collective 'power lever' for a pair of throttle-like thrust control levers like that in the Harrier. This control layout was intended to simplify the transition of both helicopter and fixed-wing pilots to the tiltrotor and make operation easy in all

Top: **The first aircraft on the occasion of its first public flight, flown for the news media several weeks after the first flight on March 19.** Jay Miller

Left: **Aircraft 2 is seen on its arrival at Putuxent River NAS on 24 July 1990. It was the first Osprey to land at the site of most V-22 flight testing in the coming years. Note the slight forward tilt of the nacelles for taxi.** Author's collection

Right: **A very new and clean ship 3 performs some of the earliest external load testing from the aft hook. Noteworthy is the angle of the open hook access doors.** Bell Helicopter

Bottom: **Aircraft 3 became the second Osprey to land aboard ship, although the vessel was not underway.** Bell Helicopter

modes of flight, and also more suitable for incorporating control switches. Another change from the XV-15 was to make the nacelle rotation switch a proportional rate controller, with the amount of switch displacement determining percentage of full rotation rate commanded.

The USMC had initially expected to transition many Harrier pilots to the Osprey. One vital difference was that the AV-8 spent little time in hover and transition while the V-22 would operate much more like a helicopter. An argument that the helicopter mode, near the ground, was the most prone to mishaps and that a collective was best suited to this mode, with simulator trials appearing to validate this conclusion, did not carry the day. A primary proponent of the TCL was Colonel Harry Blot, then V-22 program manager and a Harrier pilot. Consequently, the TCL was a feature urged on the development team by their customer and eventually became part of the specification. The decision was controversial and some wryly called the TCL the 'blottle'.

The peculiarities of the Osprey, altering fundamental control technique depending on the flight mode, raised the important issue of training. Should a tiltrotor pilot be trained initially in helicopters or airplanes? Hover skills are the most valuable to be drawn from helicopters, although the MV-22 mission would require it to hover less than 10% of its flight time. However, the use of collective and other helicopter flight techniques and characteristics would likely instill habits that would have to be 'unlearned' during tiltrotor training.

The elimination of the collective in favor of the TCL is a subject likely to be debated between helicopter pilots learning to fly the aircraft for decades to come. Helicopter pilots 'pull power' (raise the collective) to increase rotor lift if the aircraft is sinking. In the V-22 the pilot must 'push' the TCL for the same result. More than once in the years to come a V-22 pilot reverted to 'helicopter thinking' when an unexpected sink was sensed and instinctively pulled the TCL back – tightening muscles to drawing the forearm back as with raising a collective – and a hard landing resulted. Use of motion-based simulators to replicate the sinking sensation should help to break this habit. One approach is to think of the Osprey as an airplane that hovers rather than as a helicopter.

Bell's rotor hub for the XV-15 provided the starting point for the V-22 design. This was also gimbaled, although with an elastomeric gimbal hinge vice the XV-15's metal hinge. Large out-of-plane flapping was again a characteristic, with no hinges for in-plane flapping. The tilt axis cross-shaft common to the XV-3 and XV-15 was deleted in the V-22 in favor of electronic synchronization using highly reliable and redundant components.

The T406 was the largest engine ever used on a rotorcraft, with substantive changes made for the V-22 application. These included going to a two-shaft free turbine, installing five rows of variable stators in the compressor section to help prevent compressor stalls, adopting annular combustor components from the Allison T701/570 engine to replace the heavier T56 cannular design, and ensuring adequate lubrication with the engine oriented vertically. Advanced features included single-crystal high-pressure turbine blades, new rotor dynamics and turbine design, and an improved high-efficiency compressor. The powerful engine required IR suppression to reduce aircraft signature, and the V-22 was the first to feature a IRS integral with the nacelle. This feature alone cost the engine 6% of its performance.

The contractor team performed comprehensive engineering analysis and laboratory testing during the detailed design process. Additional wind tunnel and rotor whirl stand work brought the total to over 9,000 hours of such testing. Tunnel data and computer modeling proved essential in dealing with the engine nacelle inlet that had to be carefully designed for adequate air mass flow with acceptable distortion for all nacelle orientations to the incident airflow.

The V-22 design incorporated the unique pivoting wing to reduce the aircraft's dimensions for ship deck movement and stowage. It was to take up no more room in a stowed configuration than a CH-53. Fully automated, this involved folding the proprotor blades over the wing, rotating the engine nacelles level, and then pivoting the wing until parallel with the fuselage. This blade fold/wing stow (BFWS) feature was a challenge to design, especially as it had to accommodate hydraulic and fuel lines, and over 1,900 electrical connections through the rotating interface. The BFWS dictated a

Left: **Aircraft 3 became the second V-22 to land aboard ship (the Wasp), although the vessel was not underway.** Jay Miller Collection

Bottom: **Aircraft 1 on the ramp at Arlington Municipal Airport and immediately prior to its second flight. Noteworthy is the fresh paint, the missing test equipment data relay lines, and the addition of the propeller spinner caps...which were absent during the first flight.** Jay Miller

three-blade rotor system to facilitate the folding, but there were also advantages in this choice from a dynamic stability, weight and mechanical simplicity perspective. The stow rotation ring and accessories under the wing center gave a distinct hump, earning the aircraft the nickname Quasimodo.

Aerial refueling would be essential for many missions and so a fixed AR probe was installed in the lower starboard nose. An initial concept had a 'dogleg' probe issuing from the top of the nose just ahead of the center of the windscreen. This was probably deemed unacceptable because of proximity to the CV-22's radar pod and because of likely fuel spray onto the windscreen. An early concept drawing showed an external fuel tank mounted off each side sponson, but it was found better to simply expand the volume of the sponsons for greater internal capacity. Self-deployment was to be possible with one or two long-range tanks installed in the cabin.

For the USAF mission the Texas Instruments AN/APQ-174B radar was selected to permit TF/TA flight. The unit was to be housed in a small pod and radome installed in the port side of the nose. The Hughes AN/AAQ-16 Infrared Detection Set provided the FLIR camera to aid in night operations, installed in a ball turret under the nose.

Initially the Bell Boeing V-22 design had a simply tapered aft body like the XV-15. But, this was changed to one resembling the C-130 Hercules' abrupt up-sweep to reduce drag. It also improved ceiling clearances during loading of tall cargo via the rear loading ramp, and increased clearance between the top of the aft fuselage and the starboard engine nacelle during BFWS. The kink in the aft-most portion of the fuselage contributed to all of this, but was most valuable for increasing tail clearance during loading on the ramp. This eased the roll-on/roll-off capability envisioned for wheeled vehicles and engine transport dollies. The enlarged sponsons were actually found to reduce drag, especially in combination with the new aft body design.

The internal floor had to accommodate two 54x88-inch 463L half pallets or four standard 40x48-inch cargo pallets. The fuselage was lengthened from that in the proposal for greater internal volume, with dimensions equal to or larger than the CH-46. A full-scale mockup of the fuselage was constructed to verify fit and function of components and subsystems, and a mockup of the tilting nacelle with the blade fold was fabricated to demonstrate these features.

Wind tunnel testing yielded the best design of the flaperons to reduce download and thrust loss. Download lift losses of 10.1% were eventually accepted. Analysis, wind tunnel, and later on-aircraft ground vibration trials soon provided a design that negated aeroelastic instabilities. The aerodynamic tests optimized the configuration and sought to reduce rotor- and airflow-induced vibration.

The Osprey was planned to become the first combat aircraft to have nearly all of its primary structure built of composites. Initially limited only to the wing, fuselage and tail, the effort to cut weight, increase crashworthiness, survivability and water flotation saw the nacelles and proprotors also become composite components. The change to composite proprotors was more conducive to forming highly twisted

Right: **At one point during the program aircraft 1 was given international orange fields to improve visibility of the all-white aircraft. Here it is seen during the final moments of gear retraction.** Jay Miller

Below: **A low-pass by ship 1 in airplane mode reveals the orange panels on the bottom of the nacelles and wings.** Jay Miller Collection

blades and would not shatter into high velocity 'shrapnel' during an accidental blade strike. Yet the demands on the structure remained high. The cockpit enclosure had to protect the crew from an impact with a 3.0 lb (1.4kg) bird at 275kts forward velocity. The ability to tailor a structure's dynamic response by orienting composite fiber layers within the laminate was quite helpful. Even such highly loaded components as the interconnected drive shaft and the nacelles tilt axis spindles were made of composites.

By weight, the airframe was made up of 59% carbonfiber, 10% glassfiber, 11% other materials, and just 20% metal. More than 70% of the structural airframe was composed of composite materials, yielding as much as 25% weight savings. Metals were used where their particular properties justified the weight. For example, the only major metallic structural member in the wing was the outboard ribs where high concentrated loads from the nacelle attachment and pivoting mechanism would be experienced. Manufacturing time and cost were reduced through the use of composites while also simplifying assembly by virtue of the dramatic reduction in parts. The primary components featured co-cured skins, ribs, stringers and caps. For example, the complete wing upper and lower wing sections, and the belly skin and internal structure, were laid-up as single pieces and then cured in a huge autoclave.

The V-22 manufacturers were among the first to attempt mass-production of large composite airframe structures and there was initially a challenging learning curve. Concerns were raised about ballistic tolerance and survivability from enemy fire, and the rapid repairability of the composite structure under combat conditions. The aircraft was required to continue flying for five hours and land safely with structural damage. Considerable analysis and laboratory testing was performed to address these issues. A field repair kit and training program was devised. Initial ballistic impact trials of the structure were performed at China Lake Naval Weapons Center. This included firing a 23-mm anti-aircraft round into a sample wing with fuel tank under a simulated 4-g flight loading. The round cause only localized damage while the structure continued to carry its load. Similar testing also afforded the opportunity to check the effectiveness of the Halon fire suppressant system. The resulting fire was extinguished in just 40 milliseconds. The Osprey was one of the first aircraft to be held to strict combat survivability criteria. Despite these efforts, the extensive use of composites in the combat aircraft would remain an issue exploited by Osprey opponents throughout the program, and was likely to remain a concern until this or another such aircraft survived significant battle damage.

In one of the many 'firsts' of the V-22, Computer Aided Design was used extensively during the detailed design. This helped to ensure that the parts designed and manufactured thousands of miles apart by different companies would join with little difficulty. Bell and Boeing soon had 1,500 other subcontractors, vendors and suppliers in 47 states and a few foreign nations contributing to the Osprey. Apart from seeking the best resources, spreading the funding across as many states as possible helped secure broad political support.

The first wing structure, to be used on the GTA, was completed in April 1987. Construction of the wing of the first flight-worthy V-22 began in June at Bell. The first fuselage was completed at the Boeing plant in Ridley Park, near Philadelphia ('Philly'), on January 1988 and flown to Texas via a C-5A. Within days the wing had been mated to the fuselage. The engine nacelles were installed during February. The 'smoke test', during which the hydraulic and electrical systems are powered for the first time, was performed in April. This saw the engines tilted through their full arc, the ramp operated, and the nose gear cycled. Assembly was essentially completed in May and first flight was expected during the summer with initial deliveries by still set for late 1991.

By the beginning of 1988 the manufacturers were already aware that the aircraft would exceed the guaranteed empty weight and redoubled their weight reduction efforts. Weight growth is not unusual during aircraft development and expectations were that the V-22 would be at least 1,200 lb (544kg) heavy, with an associated impact to specified performance. Allison suggested that it could increase the T406 engine's power output to 7,400shp from 6,150shp, but the Navy resisted changes to this aspect of the program. It was expected that flight test would reveal that structural weight savings could be realized without sacrificing strength.

Ground testing played a vital role in verifying design choices and optimizing systems preparatory to first flight. Facilitating this was the Flight Control System Integration Rig, a

functional 'iron bird' built in Philadelphia using actual aircraft hardware with representative dimensions and masses. Avionics development was performed at the Boeing Military Aircraft Company in Wichita, Kansas, although eventually moved to Philly. There the Avionics System Integration Lab permitted testing and trouble-shooting of the hardware and supporting software. Software eventually exceeded a million lines of code; then an extraordinary figure. Boeing's hot bench tests of the integrated systems commenced in 1987. Electrical systems bench tests helped to finalize the design of this vital aspect of the aircraft. A Flight Simulation Lab provided a high-fidelity simulation model of the Osprey with a cockpit and out-of-window visual display. It was used for flight control law development and flying qualities assessment by pilots. More than 1,000 hours of motion-based manned simulations were performed to optimize the flight controls, displays, and cockpit lighting, and to train flight test crews. These three labs were eventually linked for hardware-in-the-loop piloted simulations and to check system interfaces, helping resolve problems and tune performance prior to flight test where such work would be more costly and time consuming.

Bell performed vital interconnect drive system, drive shaft coupling, and proprotor gearbox bench tests to support thousands of hours duration, limit condition of 8,600shp, and failure mode evaluations. A fuel system test rig had tanks and system components installed in a framework such as they would be in the actual aircraft, and suspended from a gimbal so that the assembly could be pitched and rolled.

Testing with the GTA began in late September 1988. The GTA had an entire wing, with the engines and full drive system, mounted on a wheeled framework. This permitted drive system endurance, high-risk whirl, and aeroelastic stability tests to be performed safely. It was operated through full conversion on a new runstand at Bell's Flight Research Center, Plant 6, at the Arlington Municipal Airport, Texas. A pilot operated the GTA from a nearby blockhouse. The limited evaluation of vital aircraft systems and rotor performance helped to uncover and resolve problems before encountering them in flight. The rig would ultimately complete 249 hours of operation that included maximum power runs exceeding the takeoff rating of the drive system. The final use was BFWS endurance trials.

Another runstand was built later as more aircraft entered testing. The V-22 could be lifted via pads under the wheels and then lowered onto jacks so that the gear could be retracted. The runstands, replete with electrical, lighting, fuel, avionics cooling, and fire fighting services, were vital in allowing systems to be wrung out, with full tilting of the nacelles and under full power, prior to flight.

The STA permitted more extensive verification that the airframe could withstand simulated ground, flight and landing loads. The 1.5-year static loads testing at Boeing also supported first-flight clearance. The static loads article was later subjected to 'drop' tests of the entire airframe to 12fps (3.7mps) in about 200 drops, followed by a drop simulating a maximum 14.7fps (4.48mps) sink rate. This demonstrated the ability of the airframe to react the loads from high sink rate landing impacts, and certify the structure and systems sound for shipboard

Facing page, top: **Boeing assembled and flight tested the Osprey at its New Castle County Airport facility, Wilmington, Delaware. The first flight of ship 4, on 21 December 1989, is captured here.** Author's collection

Facing page, bottom: **The first formation flight of Ospreys included aircraft 1 and 2 on 3 November 1989 near Arlington. Aboard the two machines were test pilots Dick Balzer, Dorman Cannon, Ray Dunn, and Roy Hopkins.** Bell Helicopter

Right: **A bright and clean ship 4 is shown in cruise with insignia orange panels added to improve visual tracking of the camouflaged aircraft.** Author's collection

Below: **Aircraft 1 approaches the Bell Helicopter Plant 6 flight test facility at the Arlington Municipal Airport, Texas. The large building and the second runstand were erected to support the V-22 program.** Bell Helicopter

Left: **Because of heating issues unrelated to the shipboard trials, the IR suppressors had been removed from aircraft 3 before it went to sea, exposing the engine exhausts. Beyond is a folded ship 4 and a CH-46 'chase plane'.** Jay Miller Collection

Bottom: **During the first shipboard trials in early December 1990, aircraft 4 was folded and moved about the flight and hangar decks to evaluate clearance and ease of transportation. The aircraft is seen here on the hangar deck under tow.** Bell Helicopter

deck landings. The landing gear was separately subjected to strenuous testing in a special test rig to impact loads of 24fps (7mps), essentially representing a crash landing condition. Fatigue testing at Bell simulated 10,000 flight hours of specific loading conditions and 30,000 landings for two airframe lifetimes, analytically extending this to four lifetimes. No substantial structural failures or redesigns resulted from any of this work.

An Ice Protection System (IPS) was designed to allow the aircraft to fly into known moderate icing conditions. Preliminary propulsion system icing tests were performed in a wind tunnel to provide an operational demonstration of engine inlet and proprotor blade deicing, the engine's insensitivity to shed ice ingestion, and allowed optimization of the IPS. Model tests in the tunnel with ice shapes demonstrated that planned vertical tail de-icing was unnecessary. Another part of the build-up to all-weather clearance was lightning effects testing.

The ability of the V-22 to stay afloat and upright long enough for safe personnel egress following a ditching at sea was evaluated with a 1/12th-scale model of the Osprey. This was equipped with powered proprotors, and a variable weight and center of gravity (cg) capability. Subjected to various simulated wind and wave magnitudes, the results showed that the basic aircraft possessed the desired characteristic without the need for flotation aids.

Flight Test

The MV-22 and CV-22 flight test program was planned to encompass 4,110 flight hours and run a little less than four years through March 1993. It would be a longer and more comprehensive test program than undertaken for any other rotorcraft. The primary reason for this was the multiple missions, the complex integrated systems, and the multiple flight modes. Most of the flights would be the responsibility of the contractors and, with the teaming arrangement, flown at two widely separated sites. These were Bell's Plant 6 and Boeing's flight test center at the New Castle County Airport, Wilmington, Delaware. Anticipating the V-22 workload, Bell constructed an 80,000ft² (7,432m²) expansion to its Arlington facility.

The testing at separate contractor facilities was a cost-savings change from initial plans to bring the aircraft to NAS Patuxent River, Naval Air Test Center, Maryland, early on. Revised plans called for 61% of the testing to be performed by the contractors and 39% by the military. The latter were to be logged by combined contractor-military crews at 'Pax' River to allow an ongoing assessment of military utility during the course of development. The contractors would perform all the high risk testing and envelope expansion work while the government looked at performance versus requirements and mission suitability. A few off-site tests were planned, including sea trials aboard ship. Flights by military crews alone would be performed during operational evaluation (OPEVAL) at a number of locations. Motion-based manned simulators and the XV-15 were employed for aircrew raining.

The first V-22 was revealed to the world in a rollout ceremony on 23 May 1988 at Arlington. At that time the first flight was scheduled for June (soon changed to around 15 August) and initial delivery of the first production example in December 1991. Flight test slipped and the contractors were under pressure from the Navy to fly before the end of November to help head off potential budget cuts. A new first flight date of mid-December 1988 was later announced, but this too was missed. Contributing to the

After aircraft 1 was returned to its red and white test colors and 5 was lost in an accident, ship 4 was the only Osprey left in the Marine colors. It is seen here aboard the USS *Wasp* in December 1990 folded and shackled to the deck. DoD

delays were GTA problems that were slowing essential gearbox tests and delaminating proprotor blade grips that had to be replaced.

Initial engine ground runs with aircraft 1 were not completed until 15 August and runs preparatory to first flight commenced on 28 December. The GTA work allowed the aircraft ground runs to be reduced, but No 1 still required 54.8 hours on the runstand. This included a full 'shakedown' of the drive and rotor system, and flight controls. Electromagnetic compatibility testing was performed during this period as well, verifying that operation of aircraft systems and external emitters did not adversely affect ship functions. The new year saw only further delays. Resolving functionality and FBW controls integration issues, late delivery of flight software, and completing fatigue ground tests were principal hindrances to progress.

Some system problems were uncovered during the ground test that had to be resolved. A pilot tendency to unintentionally couple with and exacerbate the aircraft's natural lateral rocking motion on the ground was revealed in early February 1989. The response was lowly damped even after the pilot released the stick, so a stick rebalancing was required. Two inadvertent right hand engine shutdowns were experienced in mid-March that were caused by a fuel valve operating improperly. A more serious setback was suffered on 12 March 1989 when a small fire erupted in the IRS of the right hand nacelle during an engine run on the runstand. The cause was fuel, erroneously pumped to the engine by the FADECs after shutdown, pooling in the hot exhaust. The fire was quickly extinguished with only minor damage. The program remained behind schedule and pressure on the team mounted.

The maiden flight of a V-22 was finally made from Plant 6 with aircraft 1 on a Sunday morning, 19 March 1989. Boeing's Dick Balzer and Bell's Dorman Cannon performed 12 minutes of initial helicopter mode tests with hovering, pedal turns, out of ground effect accelerations to 20kts and decelerations at a height of 30ft (9.1m), and two run-on landings. The aircraft was ready for another flight the next day – remarkable for a new test aircraft – but high winds delayed the second flight until the 21st.

Envelope expansion proceeded apace and on 6 September flight with nacelles at 45° were performed. Just days later, on the 14th, Cannon and Bell's Roy Hopkins executed the first full conversion at 155kts and 6,000ft (1,829m) altitude. An airspeed of 250kts was clocked in October and 349kts in August 1990. Testing included SEO operations at 6,150shp and air restarts of the engine. The unusual aircraft was chased with either a Bell-owned Model 214 helicopter or a leased Cessna Citation II fixed wing aircraft.

The table (below) shows the tasks to which each of the test aircraft was to be devoted. Bell was to perform the basic envelope expansion and Boeing was to conduct the specific systems tests. Aircraft 4 was to get the long-range fuel tanks in the wing and provisions for TF/TA radar to support the USAF mission. As the testing progresses through 1990 it appeared that aircraft 6 would not be essential and could serve as a spare.

At the time of the rollout it was expected that the other aircraft would follow six to eight weeks after each other and all aircraft were to fly by the end of 1989. The low-rate initial production (LRIP) decision was expected in December 1991 and the full-rate production (FRP) decision in December 1993. With the late start, flight testing was reprogrammed to run through July 1994. These plans were disrupted because of program difficulties and an apparent intentional slow-down of test aircraft assembly as controversy over the V-22 arose.

Boeing was somewhat hampered by their lack of a run stand. For this reason, and to concentrate the earliest testing in one location, aircraft 2 was constructed and initially flown at Bell instead of Boeing, as originally planned. Aircraft 2 made its first flight on 9 August 1989 and was ferried to Wilmington on 6 May 1990, the first cross-country flight for the V-22 that spanned 1,210nm. No 4 flew from Boeing's Wilmington test facility on 21 December 1989. No 3 was completed and flown in Texas on 9 May 1990 while 5 was completed in Delaware and 6 built in Texas. By 5 October 1990 aircraft 1 had flown 87 flights in 69.3 hours, 2 had 72 flights in 97.9 hours, 3 had 15 in 10.2, and 4 with 34 in 37.1 hours.

Aircraft 1 through 3 were fitted with Martin-Baker zero-zero ejection seats because of the hazardous nature of their testing. However, the baseline V-22 design did not include ejection seats. Gross weight and cg conditions were achieved through a combination of lead ballast weights installed in the nose and cargo compartment, fuel distribution, and a large steel tank that could be rolled into the cargo compartment and filled with up to 5,280 lb (2,395kg) of water. The water could be dumped out the aft external load hook opening.

The program addressed and overcame a number of engineering challenges. Consequently, changes introduced during flight test, both planned and unplanned, were frequently introduced to the test aircraft during the course of the program. Typical development problems and optimization of the engines and related systems were successfully addressed. The

FSD Test Aircraft

Aircraft No.	Principal Operator	Projected Flight Hours	Planned Principal Purpose
1	Bell	615	Envelope expansion (loads, high AOA, flutter, vibration), high altitude and H-V performance, heavy weight takeoff and landing
2	Boeing	650	Flight control system and flying qualities, icing, aircrew training
3	Bell	565	envelope expansion, flight loads and structures, vibration and acoustics, initial sea trials
4	Boeing	1,080	Proprotor/propulsion, performance, avionics, shipboard compatibility, BFWS, climatic laboratory, USAF variant evaluation
5	Boeing	610	Avionics integration, autopilot-coupled flight controls, aircrew training, operational evaluation (USMC roles)
6	Bell	590	Mission equipment demonstration, electromagnetic environment, icing, operational evaluation (USN/USMC roles)

Top: **The Osprey fit on the elevator and through the hangar doors the first time without difficulty. Aircraft 4 is shown aboard the USS *Wasp* (LHD-1).** Bell Helicopter

FSD aircraft had three pitot probes installed ahead of the windscreen and three flush static ports on each side of the nose. This layout was found to be subject to sensing errors of unacceptable magnitude because of aerodynamic interference, and the three pitot probes were felt to be so close together that a single birdstrike could disable all three. The hardware was replaced with four carefully shaped 'dogleg' pitot-static probes, two on either side of the nose. Air scoops were installed in the nacelles after it was found the nacelles and proprotor gearboxes lacked adequate cooling airflow, especially above 60° tilt.

Work quickly identified performance shortfalls requiring concentrated effort to quantify the lift and drag, finding areas of flow separation and then performing aerodynamic 'clean-up'. One change to result were vortex generators added across the midwing fairing. After trying solutions on aircraft 1 that included nacelle strakes, a wing fence was added near each nacelle to redirect vortices shed from this region that otherwise generated empennage buffet. Large openings in the nose gear doors were covered.

Fixed surface rebalancing to 'detune' the structural modes, the wing fences, rotor rebalancing, and the addition of pendulum absorbers in the hubs to damped rotor-driven vibrations all failed to adequately quell the vibrations to the satisfaction of the pilots or specifications. A vibration suppression system (VSS) system based on hydraulically driven masses was introduced below the cockpit floor of the test aircraft. This was an electrically controlled system that sought to actively reduce specific vibration modes by moving the masses in an opposing motion. Concentrating on the cockpit helped reduce pilot fatigue and vibration-induced failures of sensitive equipment in the forward fuselage. Although reducing vibration levels by about four-fifths, comparable to a turboprop aircraft and below specification limits, the VSS never quite worked properly.

The tiltrotor has always presented additional challenges with regard to vibration and the greater likelihood of rotor and structural modes coupling to create new problems, especially in airplane mode. But, the electronic flight control system meant additional challenges. The resonant frequencies were picked-up by rate gyros and accelerometers and fed back through the system. Although filtered to some extent, these still created instances of instability, both in flight and on the ground, and with or without the pilot inadvertently contributing. The desire to have a highly responsive aircraft, despite its missions as a transport aircraft, led to high gain features of the system that were found to increase susceptible to instability. The FBW also provided the most ready means of resolving the problem (save for mechanical control modifications) through gain changes, modifying or 'shaping' the pilot inputs, and the adjustment of electronic filters. More common resonances, control sensitivities, and handling qualities improvements were addressed in the same manner. Dealing with a largely 'electric airplane', many problems in other systems could also be resolved through software changes.

High workload in hover was attributed to the IGE lateral darting or skittishness and an unstable rolling tendency, both previously experienced on the XV-3. The task of controlling pitch attitude was eased with automatic flap modulation as a function of airspeed and nacelle angle. A pitch coupling with sideslip in quartering headwind was also uncovered. Pitch-up with sideslip (PU/SS) was a consequence of the rotor wake being blown onto the horizontal tail and producing a pitch-up tendency. At its worse the pilot could run out of longitudinal cyclic and trim while compensating, making it impossible to react to gusts or other unanticipated inputs. With work the V-22 achieved excellent flying qualities with all flight control features functioning and acceptable handling qualities in failure states. The author, with predominately light aircraft experience, was able to fly the V-22 simulator in all modes with no difficulty.

The rotor wash from the V-22 was stronger than a conventional helicopter of similar weight but comparable to that of a heavy lift helicopter. Repeated demonstrations of operations with ground personnel and gear beneath the hovering Osprey showed that it was possible to work safely in this area. The concept of operations would probably have personnel enter and exit the machine via the rear cargo ramp with engines running to avoid the strongest ground flow.

The Osprey's especially hot engine exhaust was another area of concern. The IR signature of the V-22 proved to be in excess of stringent requirements dictated for survivability, and the original IRS generated excessive drag. The Osprey's signature was actually a sixth that of a medium-lift helicopter and the lowest of any aircraft then in the DoD inventory. When close to or on the ground in helicopter mode the hot exhaust plumes flowing inboard was uncomfortable for ground crew working around the aircraft and it raised the fuselage skin temperature. More significantly, the hot air was drawn into the fuselage-mounted avionics cooling air inlets, taxing the cooling system. Design changes would be required.

The aft nacelle was modified as a variable geometry suppressor exit consisting of two titanium panels on the inboard and outboard sides that were electrically operated. On and near the ground these deflected outboard to turn the flow away from the fuselage. Up and away they opened and closed to vary the exhaust area at the end of the nacelle for optimal signature. Automatic positioning of the doors resulted in full-open during VSTOL and partially open during APLN. In APLN, the pilot had the option of closing the doors to three-quarters full to maximize the small percentage of engine exhaust thrust (about 2.5% of aircraft weight). During testing an 8in (20cm) extension was installed on the inboard IR suppressor door to enhance the outboard deflection of the exhaust. While the new IRS functioned as expected, the additional weight and complexity was not welcome and efforts continued to find alternatives. There were also configuration changes from the effort to reduce the external drag of the IRS.

Owing to mounting geometry on the forward-swept wing, the nacelles are canted or 'toed' outboard about 1.8° when upright. Under power the proprotors generate an upward bending moment on the wing that reduces the toe-out, but otherwise the angle slightly lowers

Left: **One solution to the airframe heating problem was to mechanically direct the engine exhaust away from the fuselage when on the ground with the deflector panels seen here.** Jay Miller Collection

Right: **Ship 5 met an ignominious end on its first flight, 11 June 1991, when miswired roll rate gyros rendered it uncontrollable. Note the paint pattern on the bottom of the aircraft.** Author's collection

the hovering vertical thrust. An attempt to recover the lost lift was Opposed Lateral Cyclic (OLC) that generated inboard tilt of both rotor disks at airspeeds below 40kts. OLC also reduced lift loss through recirculation and lessened the strength of the groundwash beneath the aircraft during hover or while on the ground, easing movement of ground personnel and alleviating some of the avionics heat load. An evaluation yielded 0-4° selectable tilt for further flight testing.

Other unique flight control modes were introduced for evaluation. Like the XV-15, a Lateral Translation Mode was provided with an associated thumbwheel on the TCL. This commanded cyclic action to symmetrically tilt the rotor disks up to 8° in such a fashion that allowed sideward translation or hover in a crosswind while maintaining a level fuselage deck. Alternatively, LTM could allow the deck to be tilted up to 8° in a hover to more safely execute a cross-slope landing, reducing pilot workload and providing more clearance under the upslope nacelle. However, LTM risked damage to the elastomeric rotor hub springs if inadvertently engaged during ground operations (as demonstrated on a few occasions) and this was felt to outweigh the advantages. It also created disturbing side forces on the crew, especially for the crew chief who might not be seated.

LTM was eventually dropped, but Lateral Swashplate Gearing offered similar benefits without direct pilot control. It reduced the bank angle required to maneuver in lateral translation at low airspeeds by commanding a bit of lateral swashplate tilt in addition to the differential collective. It worked together with Differential Collective Pitch for optimal roll control power.

A selectable Precision Flight Mode changed lateral cyclic from a roll rate controller to a roll attitude controller for such operations as aerial refueling and formation flight. Augmenting this was a feature active in APLN only that caused the rotor to over-speed briefly when reducing power, or caused rpm to 'droop' temporarily when the power was advanced. This enhanced precision speed control for the exacting flight tasks.

As pressure mounted on the program, the testing was accelerated in 1990. This was meant to quickly demonstrate the great potential and soundness of the tiltrotor technology, but also provide data for validation of reliability, maintainability and survivability data. The results would hopefully win converts in Congress. The Navy also suggested that the flight test program could be reasonably reduced to 3,000 flight hours. However, high flight test productivity proved elusive, averaging only 8 flight hours per month per aircraft vice the targeted 15-25. Contributing to this was low reliability of the pre-production airframes, the many development problems encountered and corrected, the frequent maintenance periods, and multiple changes in test configuration.

The government side of the flight test program was divided into several distinct periods when they took over operation of one or more aircraft to perform dedicated evaluations. These were either development tests (DT) or operational tests (OT) as summarized in the table. The OT-IIA period was also intended to provide the confidence to approve LRIP, or pilot production, of 12 Ospreys. LRIP aircraft would provide the production-representative examples for a limited three-month OPEVAL that was to lead to another LRIP approval for 45 machines. The final yearlong OPEVAL was to begin at the end of 1990 (later moved to July 1991) and support a FRP decision by the end of 1991 (also slipped). The evaluation was to be conducted at Marine Corps Air Station (MCAS) Quantico and other sites by the Marine Experimental Helicopter Squadron 1 (HXM-1). Program delays disrupted these plans.

FSD Flight Test Summary

Phase	Aircraft No.	Dates
FSD	1-5	19 March 1989 - 20 July 1992 halted prematurely
DT-IIA	1 and 2	17 March - 23 April 1990
OT-IIA	1 and 2	15 May - 8 July 1990
DT-IIB	3 and 4	5 November - 14 December 1990
DT-IIC	2 and 4	2 April - 19 August 1991 halted prematurely
–	–	further evaluations canceled

Phases DT-IIA and OT-IIA was conducted at Bell and offered an opportunity for an early assessment of the V-22's mission potential. The DT portion encompassed 30 hours of tests flown by a trio of USMC pilots. By the end of this series of tests the four-aircraft fleet had about 220 flight hours. Objectionable airframe vibrations, especially at high speeds, were the dominant criticism. However, the planned VSS had yet to be installed. The overall conclusion was that the aircraft showed great potential for meeting mission requirements.

DT-IIB consisted of the Phase 1 Shipboard Compatibility Trials. The land-based phase, preceded by BFWS demonstrations, commenced in November with electromagnetic compatibility (EMC) ground testing and 11.7 hours of training over 10 flights. Aircraft 3 and 4 were flown out to the USS *Wasp* (LHD-1) during 4-7 December 1990. The V-22s staged out of Pax River to the vessel 50nm off shore and No 4 became the first Osprey to land aboard a ship with test pilots Dick Balzer and Major Gerald Hammes, USMC. Aircraft movement on the flight deck, elevators, and hangar deck, supportability trials, and human factors work were conducted. The shipboard testing spanned 5.2 flight hours and 5 sorties, and included 15 takeoff and landings from various spots on the deck using a number of approach and departure patterns.

Sea trials are important because of variations in wind-over-deck angle and velocity conditions at various spots due to airflow off the vessel's superstructure plus other aircraft and vehicles on the deck. The differences in these conditions and the ship's motion due to sea state can create challenging takeoff and landing conditions for rotorwing aircraft. However, at this early stage of V-22 development the variables were reduced by having the ship at a standstill. Also, because the EMC testing had apparently found areas of concern or was not completed, all emitters on the *Wasp* were turned off to avoid possible interference. Insights into visibility from the cockpit and special procedures, such as using nacelle movement to make positive changes to closure rate on final approach, were essential lessons. Downwash effect on deck personnel and operations were found to be negligible or similar to other helicopters. One advantage of the small deck of the amphibious ship was that, with one proprotor over the edge of the deck, the ground jets from the two proprotors did not join and produce a strong outwash forward and aft of the aircraft. The asymmetric downwash at deck-edge did not unduly increase control workload. Another concern was that hot engine exhaust would damage the deck or equipment, such as life raft containers and fuel servicing points just below the deck edge. This, too, proved to be unfounded. No significant negative results were uncovered aboard ship and further sea trials were planned for late the following year.

Typical of any developmental aircraft, government testing identified many significant and minor deficiencies. Among the problems was unacceptably high pilot workload during the IGE low-speed/hover phase of flight. The testers found inadequate mission radius to meet the USN mission and it fell 180nm short of the USAF's 700nm requirements. There were drive systems problems, an inability of the troop seats to accommodate a fully equipped Marine rifleman, and overall low system reliability. Even the company pilots judged the cockpit displays deficient in ease of use and suitability of the displayed information, with too much button-pushing for even the most routine tasks. On the plus side the excellent visibility from the cockpit was judged to facilitate shipboard operations and landing in confined areas. The potential of the design to satisfy its US military missions was readily evident.

Sling load trials began in February 1991. This commenced with a 2,000-lb (907-kg) ballast hanging from the aft hook, but this was extended to 4,000-lb (1,814-kg) and dual-hook loads were soon being lifted. The ability to take-off and accelerate to 100kts in less than 30 seconds with a sling load was phenomenal and would be of great tactical value. The Osprey would be the first fixed-wing aircraft capable of carrying a sling load. By the end of February 1991 the four test aircraft had accumulated 400.8 flight hours over 340 flights.

The purpose of DT-IIC was an intermediate technical evaluation of the V-22 in support of a DoD limited production decision and to evaluate readiness to begin the first phase of OPEVAL. The flight portion at Pax was prematurely terminated with the crash of aircraft 5 on 11 June 1991. All V-22s were grounding pending results of the accident investigation.

Aircraft 5 was just 3 minutes into its maiden flight at Wilmington when it impacted the ground from a 15-ft (4.6-m) hover. Handling difficulties in the roll axis became evident almost immediately on liftoff, with aircraft response wildly out of phase with pilot inputs. The first attempt at setting the aircraft down resulted in the left nacelle impacting the pavement and destroying the IR suppressor. In the next attempt the left nacelle and proprotor struck the ground, flipping the ship over onto its back. The aircraft slewed around within 200-300ft (61-91m) before coming to rest with the cockpit separated at the splice joint. Both pilots escaped with only minor injuries. A small fire was quickly extinguished but aircraft 5 was destroyed. The V-22 test fleet had logged over 550 hours in 463 flights to that time.

The crash of 5 was attributed to two of the three fight control system roll rate gyros being wired in reverse. A reversal of the wiring in one gyro had been identified during the manufacture of both aircraft 1 and 3, and corrected, but the potential for the mistake was not properly communicated to those assembling 5. The error was not caught and was actually made worse by a second gyro being mis-wired in that the control system voting logic would discount the proper gyro signal. Also, the flight control BIT, that would have detected the fault, was not run prior to the flight. The reversed sensing of the two gyros rendered the aircraft uncontrollable. Changes to drawings and quality control measures were implemented to help prevent such future errors. The problem was clearly not associated with the unique nature of the aircraft, although this fact was generally lost in press accounts. The remaining test aircraft were flying again by 10 September 1991.

Aircraft 4 spent February through July 1992 in climatic testing at Eglin AFB, Florida, being subjected to temperatures ranging from -65° to 125°F (-54 to 52°C), rain at up to 5in/hr (12.7cm/hr), freezing precipitation, snow, and wind to 45kts. The climatic testing succeeded in qualifying V-22 systems for extreme weather conditions while identifying areas for improvement. Field trials with the same aircraft in icing conditions were planned for later in Canada.

Aircraft 4 left Eglin on 20 July 1992 for MCAS Quantico, Virginia, where the machine was scheduled to conduct a week of troop exiting trials and operational demonstrations. However, during reconversion on the last 18 seconds of its approach the aircraft suddenly descended at high rate, rolling slightly right and nose down. It impacted the Potomac River from 100-150ft (30-46m) about half a mile from the runway. The seven aircrew and test team members were killed. The airframe had accumulated 103.4 hours of flight time in 94 flights to that date. The entire fleet had 762.6 hours in 643 flights.

The subsequent investigation revealed that flammable fluid, for which there were several potential sources, had leaked and pooled within the right nacelle near the engine inlet. This then flowed into the engine when the nacelle was tilted up, causing three power surges and a fire. The engine then failed. The backpressure from the surges damaged the inlet and engine nose cap, allowing the fire to migrate into the nacelle. The crew was by then fighting for their lives with a cascading serious of faults and system failures. The sudden power loss was not fatal, the port engine continuing to drive both proprotors, but the nacelle fire reached extreme temperatures. With flame applied directly to the composite drive shaft, the vital component was critically damaged in just seconds. When this failed the essential link between the proprotors was severed. As the rpm began to decay on the right proprotor, the left engine power was automatically reduced to a low level to prevent loss of control. This near total loss of power doomed the aircraft. Further-more, the shaft failure released hydraulic fluid and, coupled with a flight control computer electrical failure, dramatically reduced control authority.

Left: **The need to operate the aircraft in all flight modes within the large McKinley Climatic Laboratory hangar saw the airplane mounted on a raised trestle to accommodate nacelle tilt. The most complex setup the lab ever dealt with included large steel ducts called 'crab claws' that carried away the engine exhaust.** AFFTC

Opposite, top: **Aircraft 3 is seen in flight with the experimental engine exhaust deflectors and a partially open cargo ramp.** Jay Miller Collection

Opposite, bottom: **Contrary to many reports, aircraft 6 was built but never completed. It is seen at Bell's Plant 6 during timed emergency egress trials. Beyond the fuselage is probably the wing under plastic, never mated.** Jay Miller collection

The investigation found no fault with the basic tiltrotor concept or the V-22 design, but Bell Boeing made changes to the Osprey based on the findings. More drain holes were added to the nacelle where fluids might pool. Changes to the inlet helped ensure that fire would not spread elsewhere in the nacelle, and electrical ignition sources were moved. The firewall within the nacelle was extended and the cross shafting adjacent to the engine was eventually replaced with a more heat-tolerant material. In the meantime, a flame-resistant shield was installed at the outboard end of the drive shaft within the nacelle. Greater separation of critical flight hydraulic lines and control system wiring was introduced and nacelle cooling increased. Cockpit caution and warning lights for interconnect drive shaft integrity and nacelle fires were added or improved. The in-depth safety review also identified non-optimal design characteristics elsewhere in the aircraft for which other changes were introduced.

Following the fatal accident the Ospreys remained grounded for 11 months. This, political battles over the program, and the manufacturer's hesitancy to spend more of their own money without certainty of return on the investment made the V-22 appear to be dead. Bell soon ceased virtually all work on the V-22 while Boeing continued only a low-level effort. Most of the flight envelope had been cleared with the following end-point conditions achieved:

- 292kts in level flight and 349kts in a shallow dive
- 21,500ft (6,550m) altitude APLN, 15,000ft (4,570m) in CONV, 10,000ft (3,050m) in VSTOL
- 51,000 lb (23,135kg) takeoff GW, 48,100 lb (21,820kg) hover GW
- full cg range at up to 47,000 lb (21,320kg) GW
- 3.2G in APLN and 1.5G in CONV and VSTOL
- 7fps (2mps) touchdown sink rate
- 4,580shp (3,415kW) at each rotor

Trials had included initial shipboard evaluation, formation flight, night and simulated instrument tests. While many developmental issues were uncovered in the flight test, none revealed a fundamental problem with the tiltrotor concept or major aspects of the V-22 design.

The last FSD test ship, aircraft 6, was nearing completion at Plant 6. However, work was halted before June 1991. The wing and fuselage was never mated before funding ran out.

Cancellation Threat

Original plans were that the first four production lots would be bought on fixed-price contracts based on preliminary cost estimates. Delays experienced during development prompted the Navy to slip expending advanced procurement dollars for production from Fiscal Year (FY) 89 to FY90. This would delay production a year and IOC by approximately six months. Being a year latter than originally stated in the contract meant that the option of initiating pilot production at a predetermined price was lost. The plan by that time was to buy just 10 pilot production machines, with 4 planned for FY92 and 6 in FY93. There were still many voices that insisted starting pilot production was premature given the little testing that had been performed to date on a radically different aircraft with uncertainty about its suitability for the USMC mission. The overall buy of aircraft was also to be stretched out. The Osprey was to be produced in ten lots, with the largest being 132 aircraft in 1996's Lot 4, or 11 aircraft per month.

Top: **The FSD Static Test Article fuselage is essentially as complete as it gets for the purpose of its ground tests. Without the wing, the fuselage must be ballasted to prevent it tipping back.** Author's collection

Middle: **This dramatic photo captures EMD test aircraft 7 in airplane mode, following modification to support CV-22 flight testing, against an angry sky.** Bell Helicopter

Bottom: **A welcome sight for a squad of Marines eager to be extracted from a hostile area.** Ron Culp

The contractors began to receive money in February 1989 for long-lead procurement of parts to support pilot production. Congress provided the funding despite the fact that the Pentagon had not called for the money in its budget. It was then announced on 19 April 1989 that the 1990 defense budget requested no V-22 money, effectively terminating the program. Secretary of Defense Dick Cheney issued instructions on 1 December 1989 that all contracts associated with V-22 production were to be suspended. The Navy immediately canceled the $328.8 million in long-lead contracts.

The DoD's stated rationale for canceling the V-22 was that the machine was overweight and would require costly redesign and testing to meet requirements. Consequently, the anticipated final unit cost could not be justified when compared with alternative helicopters (as yet unidentified). However, coming little over a month after the Osprey's first flight, the reasons behind the decision were more complex.

The recent end of the Cold War considerably eased East-West tensions and generated pressure to reduce defense spending. President Bush had instructed Cheney to find $10 billion in cuts from the FY90 budget. Since the V-22 program had yet to enter production, it could be terminated with comparatively little loss of investment. Doubts prevailed about the Osprey's value in the post-Cold War era. Support for the V-22 within the armed forces also appeared to be slipping, with the Army having withdrawn and the Navy failing to make any substantive HV-22 commitment. The SV-22 faded in 1989 as the V-22 ran into trouble. The increasing weight of the aircraft would have proven problematic for some of the ships that would have hosted the sub-hunter. The deep cuts in the number of aircraft undermined the multi-service character of the program that had figured prominently in its justification. The reduction in the number to be procured also contributed to an increase in projected unit cost from $16 million to $34.5 million, with much uncertainty and competing figures offered by all parties.

Near the end of 1990 the contractors had exceeded the cost ceiling of their fixed price contract and were taking losses in excess of $150 million on the program, eventually ballooning to $300 million. But, they were willing to press ahead. A fixed-price contract for a complex development program with many uncertain elements was ill advised. All such contracts for major weapon systems would eventually be cancelled save for the C-17A airlifter. The partnership of Bell and Boeing, plus Allison and P&W, was not entirely efficient. While they were working closely on development and flight test, they were separately developing competitive bids and production capacity for the pilot production, entailing duplicative expense.

The V-22 pushed aerospace technology in many ways, despite the original assessment that the technological risk was low and a prototype was not required. The tiltrotor concept may have been demonstrated as sound with the XV-15, but the V-22 was a considerable departure from this comparatively simple demonstrator. It brought together many new or advanced technologies for the first time in the most complex production combat rotorcraft to that time, already unusual in fundamental aspects and with quite ambitious requirements. Considering these challenges, it is remarkable how quickly the team had built and begun testing the aircraft, and how comparatively successful the design proved to be. However, these complexities meant that the development was bound to be long. The design was clearly deficient in some areas, such as being overweight and falling short of the USAF range requirement, and many areas required further maturation before a fieldable warplane resulted. The Air Force was willing to accept the range deficiency. Falling short on some requirements was perhaps to be expected in a multi-service program. The developers needed more time to complete development, but instead schedule, funding, and personnel fell into disarray.

The USMC and USAF strongly defended their need for the V-22. Flight test had shown the V-22 performance meet or exceed nearly all requirements, with plans in hand to correct deficiencies. But, as with most major weapon system programs, there were many powerful national leaders who opposed the Osprey as too expensive and complex, saw its faults as fatal, and judged it the wrong aircraft for the mission. Many in the aviation community and military also expressed grave doubts about introducing such a radical technology into a combat environment as severe as USMC operations. The crash of aircraft 4 and 5 had certainly colored many opinions, despite investigation findings. Accidents during flight testing had become rare and their lack considered the norm.

Most prospective operators welcomed a new aircraft with the range, payload and speed of the V-22. But, rumors of unacceptable downwash velocity and exhaust temperatures, coupled with the natural suspicion of such an odd machine, had its insidious affect. The scuttlebutt was that the Osprey would drown survivors in the water or knock over persons on the ground. It would set fire to LZs or disappear in a cloud of recirculated dirt and vegetation. Just the complexity of all the moving parts and the use of composites in crucial components made some nervous. Helicopter and airplane pilots alike had to see for themselves the unique advantages of the tiltrotor and how they outweighed any disadvantages when compared to what they were used to flying. The testing had only begun this process, addressing each contentious issue raised by detractors and consistently proving them wrong.

The downwash proved comparable to that produced by the heavy CH-53. Both would bowl over the unwary, but with the proper training it was possible to ensure safety of personnel working around the machines. The relatively high downwash velocity did kick up ground material, mainly to the nose and tail. Some forward obscuration was experienced, but the pilots found it acceptable. The 'brownout' from blowing dust is also not atypical of heavy lift helicopters. Light material could circulate and end up in filters and screens, and some inside the cabin. Testing had shown that in APLN the aircraft produced about a third the noise of a CH-53 and much less than a C-130. In helicopter mode it was comparable to the 53.

The IRS reduced the exhaust plume to a tolerable 390-515°F (200-270°C) at the face of the exhaust or 50-100°F (12-38°C) above ambient at the height of a man with the aircraft in a 10-ft (3-m) hover. Although vegetation immediately beneath the exhausts was burned, the material was just blown away and no fires were ever experienced. The proprotor wash mixed cool air with the exhaust.

As for the 'thousands of moving parts flying in formation', the V-22 had the same number of gearboxes but fewer transmission shafts than the successful CH-47. Furthermore, the Osprey's FWB system eliminated the mechanical control runs found in the 47. Those who operated rotorcraft did not see this complexity as a limitation. Composite components underwent rigorous testing and suspicion about their suitability would dissipate over years of successful operation.

The V-22 was very different from the CH-46 and CH-53, and would require many changes to existing concepts of operation to integrate the new machine into the Marine and USAF missions. For example, some conditions where the powerful Sea Stallion would be able to hover over an objective would require a STOL landing by the Osprey. However, other performance aspects of the V-22 were most welcome and selling features, especially its speed. Most operators were eager to get their hands on the Osprey.

Aircraft 1 was approximately 1,000 lb (454kg) over-weight at first flight. By that point the contractors had identified changes that promised to reduce weight by 2,000 lb (907kg). However, by late 1990 the excess had reached 2,800 lb (1,270kg). Work continued through the remainder of FSD to cut this and any additional weight likely to appear as continued development revealed the need for design changes. Identified weight savings would leave some 1,200-1,600 lb (544-726kg) carried over into pilot production but eliminated in Lot 1.

An increase in the transmission rating from 4,200shp (3,134kW) to 4,570shp (3,410kW) was implemented, allowing the aircraft to use more available engine power. This would compensate for the remaining excess weight and get the CV-22 closer to its required range. The option of up-rating the engine from 6,150 (4,589kW) to as much as 10,000shp (7,462kW) was also considered, although this would require further increase in the transmission capacity. However, these options would only further exacerbate program costs.

The Pentagon's decision to cancel the Osprey appeared to undercut Congressional plans to decide the production issue in the spring of 1990 when the COEA it had ordered was to be ready. Data from the accelerated flight test fed the COEA. It concluded that the Osprey was the most cost and operationally effective choice for a broad range of missions than existing helicopters, and with lower life cycle costs. It was simply the only aircraft that met the JVX requirements. Alternatives evaluated were a mix of UH-60s and CH-53s. The report estimated that the Osprey's unique characteristics would increase survivability four-fold over the current fleet of helicopters while also improving productivity.

Although not requested by the Administration, from 1990 to 1992 substantial bipartisan Congressional support saw hundreds of millions of dollars appropriated for MV-22 and CV-22 development, testing, and building production-representative examples. Although the President signed all these DoD authorization bills, the Bush administration was not convinced that the nation needed or could afford the Osprey. The Navy released only a fraction of the money to the contractors. The service correctly stated that the funding was far less than required and the future of the program was undecided. Congress insisted DoD obligate the funds already authorized and the General Accounting Office (GAO), Congress' investigative arm, judged Secretary Cheney's orders to terminate all production contracts to have been improper. The tug-of-war between the Legislative and the Executive branches almost ended up in court.

It was difficult for the contractors to carry on activities in this uncertain climate, especially as they were spending a good deal of their own money. Manufacturing and test operations were minimized and many persons reassigned or let go. An atmosphere of pessimism and low morale was pervasive. The loss of invaluable expertise on what was still an experimental aircraft was deeply felt whenever the team resumed operations. The bursts of activity, for demonstrations instead of comprehensive testing, usually generated tremendous schedule pressures that led to unwise shortcuts of normal procedures and employee fatigue. These factors may well have contributed to the oversights leading to aircraft 5's accident and left many individuals bitter.

ARMY

Production and Service

Resurrection

The Marines continued to express their belief in and urgent need for the Osprey, making it their number one acquisition priority. The delay introduced by initiating another acquisition program would be a major impediment to modernization plans. They were convinced that another helicopter would burden them with a less suitable aircraft for decades. The tiltrotor promised twice the speed, three times the payload, and four times the range of the CH-46 – vital qualities in an era when US forces had fewer overseas bases and needed to react to distant trouble spots rapidly with overwhelming force. The Sea Knight would simply not meet Corps combat requirements in the coming decades. It and the CH-53D were approached an average 25 years age, with added flight restrictions, and becoming increasingly costly to operate.

Following the cancellation decision, a study was launched to seek alternatives satisfying the USMC needs. Over the next few years a number of individual and mixes of aircraft were considered. The DoD proposed a combination of heavy-lift CH-53Es and a derivative of the Black Hawk, called the UH-60(S) or CH-60X. The smaller size of the UH-60 would have required fundamental changes to USMC task forces, including the size of squads and the makeup of their equipment. The Corps examined upgrades and life-extension programs for the Sea Knight, tentatively called the CH-46X, to keep it operating beyond its 30-year, 10,000-hour life. Reopening production was also considered. The experimental Boeing Model 360 (BH-360) and developmental Sikorsky S-92 (not flying until December 1998) might have been revised to meet requirements. The CH-47E or a modified version (CH-47X, presumably but unlikely making it compatible with shipboard operations), in a mix with modified UH-60s was considered. Even the European EH101 and Super Puma were contemplated.

In the face of mounting acrimony between Congress and the Administration, something positive had to be done about the Osprey and the requirement. The DoD generated a number of plans delaying and stretching-out production, contributing to budget-cutting goals but preserving the program while alternatives were considered. Studies and continued Osprey FSD would, some pointed out, ensure that revising and procuring the V-22 was weighed against alternatives in seeking the most cost-effective means of meeting the need. Others saw it only as a delaying tactic while the Osprey was allowed to wither from chronically low funding. At the very least it spelled a minimum one- or two-year delay in production that would be intolerably costly to the contractors.

The Navy responded with a proposal for what was called the Medium-Lift Replacement (MLR), the notional aircraft tentatively dubbed MVX. This program was approved in 1992 for which $3.4 billion was identified in the DoD's 1990-94 five-year budget plan. The initial requirements document contained performance figures that only the Osprey could meet, and the USN was instructed to rewrite it to 'even the playing field'. The revised requirements emphasized only the short-range amphibious assault mission, setting aside for the time being the SPECOPS and SAR requirements. A 50-hr demonstration was to be held as all candidates already had flying prototypes or production examples.

The MLR requirements, when compared with the JVX, saw the cruise speed reduced to 180kts required with 200kts desired, and eliminated the 2,100-Nm self-deployment. The comparatively low airspeeds and range made it appear doubtful that anything but a helicopter would be found affordable. V-22 proponents protested that this did not emphasize the speed, range and mission flexibility of the aircraft that helped justify its higher unit cost. It also meant that further Osprey development and production would concentrate solely on the MV-22, with attendant increase in unit cost. Major revisions to the V-22 would also bring additional costs associated with reengineering and test in a vicious cycle.

Preceding page top left: **To investigate issues with refueling from the KC-135 boom-to-drogue adapter, wake surveys in the refueling envelope behind a USAF tanker was performed with aircraft 3 on 31 January 1994.** Author's collection

Preceding page top right: **Aircraft 2 flies behind a water spray rig at the end of a KC-135 refueling boom during icing trials in February and March 1994 at Pax River. Ice, colored with yellow dye, can be seen accumulating on the radar radome.** AFFTC

Preceding page bottom: **To immerse more of the airframe in the freezing cloud to test the Ice Protection System, ship 2 was flown behind the Army's modified Chinook fitted with the Helicopter Icing Spray System rig.** Jay Miller Collection

Right: **Aircraft 3 performed early rescue hoist tests during Risk Reduction trials. Note the truncated ramp and lack of cargo door for an evaluation of fastrope operations in the proposed EMD configuration.** Author's collection

Left: **Bearing the new colors for EMD, ship 2 appeared at the June 1995 Paris Air Show after having been transported to Europe by sea. The show number 122 was applied to the gear doors. Appearing beside the XV-15, the Osprey put on an impressive demonstration.** NAVAIR

Bottom: **Bottom detail and gear well interiors are visible in this shot of aircraft 2 during Risk Reduction flight trials.** Bell Helicopter

All MLR alternatives were factored into the COEA ordered by Congress, this marking the fourth (eventually reaching seven) COEAs directly associated with the V-22. Two independent mission effectiveness studies funded by Bell Boeing were also conducted during the period. This brought to 18 by that time (ultimately 19) the number of studies conducted over 25 years involving the V-22. Despite its higher unit cost the V-22 was repeatedly found to provide greater capability and combat effectiveness at overall operating costs nearly the same as alternatives. Overall life-cycle costs were expected to save billions through reliability, reduced manpower and fleet size, and greater survivability. Analyses looking at MVX alternatives to filling the USAF mission, called the Advanced Multi-Mission Lift Aircraft, yielded similar results.

While all of the studies favored the V-22, the assumptions and quantitative conclusions varied widely, and most were generated by teams with clear bias. However, with election-year political pressures and determination to resolve the matter, the decision to proceed with the Osprey was made with little dissension. Bell Boeing was offered a chance to redesign the Osprey to reduce unit cost and weight. Marine Corps IOC was reset for 2001.

The resurrection of the V-22 commenced on 22 October 1992 with an interim contract award. The $550 million funded studies aimed principally at demonstrating how the Osprey could be redesigned to meet weight and performance goals while also coming in at the anticipated unit cost with reduced technical risk. This was expected to be a lead-in to an eventual Engineering and Manufacturing Development (EMD) program. The manufacturers sought to eliminate 1,100 lb (500kg) of weight from the baseline design (2,000 lb from the FSD aircraft).

By late 1993 the DoD considered expanding the program to include the SAR and SPECOPS missions. This was formally addressed in August 1993, just weeks before a major MV-22 design review. USAF IOC was to be in 2005. It looked like requirements could expand to encompass the USAF and Navy missions without adverse impact, virtually returning to the program as it had been during FSD. Without having made the USAF mission a requirement at the beginning of the redesign, such accommodations as the uprated transmission were not included. Performance was impacted, including a reduction from the original 700nm radius mission to 500nm.

Engineering, Manufacturing and Development

The effort to meet the target empty weight of 34,182 lb (15,505kg) and reduce unit cost included a fundamental reexamination of the design and manufacturing processes. The original airframe had been criticized for excessive use of composites in areas where weight was not saved or the manufacturing process was more costly than with traditional materials. Consequently, the EMD design was to see composites reduced to 43% of the airframe. Where composites were retained, manufacturing improvements allowed a reduction in production costs, such as hand lay-ups replaced by new tape lay-up machines. The previous practice of mechanically fastening the composite skin to the composite substructure in some areas was changed to a bonding process. The number of parts and fasteners in the fuselage were reduced by more than a third. The entire design was converted to a computer-based format to more readily facilitate changes. Design and manufacturing changes to the wing stow mechanism, conversion spindles, transmission, landing gear, wing ribs, nacelles and swashplates also contributed to weight and production cost reduction. Emergency egress was altered from the traditional mechanically opened hatches to use of pyrotechnic charges. An aluminum cockpit cage replaced the titanium structure, mandating a reduction in pilot knee window size in the interest of strength. The windows are of little value in the V-22. Production quality was improved to cut waste and rejected parts. Other revisions introduced new and more effective technologies, and corrected known deficiencies.

Bell Boeing eventually exceeded their goal. By the time the first EMD aircraft had been assembled the team was 389 lb (176kg) below the weight reduction goal, giving a comfortable margin for the almost inevitable growth during development testing.

In April 1993 Bell and Boeing submitted their EMD proposal. Their redesign exercise was considered successful and the decision was made to award another $2.65 billion development contract. This included continued testing of two FSD aircraft, construction of four new flight test articles, plus six LRIP aircraft for OPEVAL. The EMD contract was signed in June 1994 with the program to run through 1998. The neglected FSD contract was canceled with nearly $2 billion expended. Under the new contract, Bell Boeing would collaborate in production using the division of responsibility established during FSD. The Navy abandoned plans for a second source for the engine.

The total cost of EMD was to be $3.4 billion. Unlike FSD, this was a cost plus biannual award fee contract, with award payments based upon performance. Congress capped the V-22 program at $1 billion per year, making cost-cutting a dominant issue. With production, the whole effort was expected to cost $37 billion.

By the time the first EMD aircraft was assembled the changes and production efficiencies got the anticipated unit cost down to $32.3 million while working to a $29.4 million goal. This assumed a 523-aircraft buy at two units per month and compared with the $41.8 million predicted in early 1993. This effort used a design-to-cost philosophy, trading capabilities when necessary to reach the cost goal. The irony would be that just a few years later, when low-rate production funding was being calculated, inflation and additions during development would see the cost back at around $36 million a copy (estimated $49.7 million for the CV-22).

In September 1994 the program was rebaselined with new target dates. Many feared delaying production further would see subcontractors and suppliers curtailing support, with costs rising to bring in replacement companies. The new plan had production funding commencing in 1996 for long-lead items and the first four LRIP articles delivered in 1997. Annual lots were to be 5 aircraft in Lot 1, 7 in each of Lots 2 and 3, and 8 in Lot 4. This would generate the 25 machines required to support OPEVAL and training. It also ensured that the fewest practical aircraft would be subjected to costly modifications to bring them to final production configuration with changes likely found necessary during testing. Full-rate production was to begin in 1999 with 9 machines. However, some still perceived these numbers as inefficiently low and uneconomical while others felt they were too many and costly prior to complete testing and a decision to proceed.

Continuing defense budget reductions created pressures to cut the V-22 program to pay for unplanned military operations and bolster readiness. Consequently, production plans were altered again. All the delivery dates were slip two years with a reduction in the total numbers. Initially the Marine requirement was for 425 aircraft, the USAF 50, and Navy 48. In 1997 the USMC quantity was reduced to 360 following a major DoD force structure review, while it was recommended the production rate increase from 24 to 36 to ensure more rapid fielding and realize savings from economy of scale.

The basic Osprey mission requirements remained unchanged, but some specifics were revised. The EMD specifications remained demanding. The 'threshold' requirements were the minimum acceptable while the customer desired the 'objective'. Some of the thresholds were identified as key performance parameters of particular interest during testing. The most salient of these critical technical indicators are given below. 'MV' indicates requirements applicable only to the MV-22, 'CV' those for the CV-22, and the rest applicable to both. The two models were expected to be 85% common (90% in hardware, 60-80% software, similarly in avionics).

The maintainability and mission readiness figures were among the most comprehensive and demanding to be adopted by any aircraft program to that time. They were important for ensuring that the Osprey could be supported with reduced manning and operations funding,

Top: **Although the V-22 requirements called for carrying a light vehicle with trailer, there was no suitable vehicle in the DoD inventory. However, an experimental vehicle was built that just fit inside the cabin – the grounded ship 2 fitted with a short ramp used for the fit check in the late 1990s.** Bell Helicopter

Right: **Ship 3 as it appeared in the summer of 2003 at the American Helicopter Museum, West Chester, Pennsylvania, much the worse for wear.** Jim Jagodzinski

EMD Specifications	Threshold	Objective
Cruise speed (3000ft alt, 91.5°F, max design GW)	MV 240kts / CV 230kts	MV 270kts / CV 250kts
Instantaneous g-loading	-1 to 3.5 APLN	
	0.5 to 3.0 VSTOL	
Mission radius		
special operations (CV)	500nm	750nm
land, troop lift (MV)	200nm	
land, external load (MV)	50nm	110nm
sea, troop lift (MV) - round trip	50nm	110nm
sea, external load (MV)	30nm	110nm
Self-deployment range	2,100nm	2,100nm
	1 aerial refueling	0 aerial refueling
TF/TA altitude (CV)	300ft (91m)	100ft (31m)
OGE hover	MV 3,000ft (914m) / CV 3,900ft (1,189m)	
Service ceiling	25,000ft (7,620m)	
STO		
shipboard	300ft (91m) with 15kts headwind	
ashore	500ft (152m) with headwind	
STO with SEO over 50ft obstacle (no payload, max continuous power, 60% fuel)	>1,000ft (305m) elevation	>7,500ft (2,286m) elevation
Accommodations		
Internal payload/cargo	MV 8,000 lb (3,629kg)	10,000 lb (4,536kg)
Internal capacity, troops	MV 24 / CV 18	CV 24
options: 12 litters, or 1 light vehicle + trailer plus 4 personnel, or 4 tandem-loaded 48"x48" platforms or 2 54"x88" 463L pallets of	4,000 lb (1,814kg) each	5,000 lb (2,268kg) each
External loads (MV)	10,000 lb (3,048kg)	15,000 lb (4,572kg) dual hook
Rescue hoist capacity (CV)	600 lb (272kg)	
Operational Considerations		
Operating temperatures -65° to 102°F (-54 to 39°C), to -20°F (-29°C) without kits		
Precision navigation	within ≤2 rotor diameter	≤1 rotor diameter
Wing fold/stow	90 sec	
Major dynamic components would operate ≥1,500 flt hours between removals (MV)		
Mission capable rate	≥82%	≥87%
Mission reliability (MV)	≥85, 3 hours mission	
Weapon system reliability (CV)	≥77%, 4 hours mission	≥84%
Launch time (CV)	60 minutes	15 minutes
Mean turn around time	≤0.25 hours	≤0.166 hours
Mean flight hours between aborts	17 hours	
Mean man hours per flight hour	≤11 hours	
Mean time between failures	1.4 hours	2 hours
Mean flight time between unscheduled maintenance	>0.7 hours	
Mean repair time	<7 hours	<5 hours
Mean corrective maintenance time	≤3.7 hours	
Minimum service life	20 yrs	30 yrs

and the number procured meeting mission demands. One significant improvement was that, instead of the daily maintenance inspection required for helicopters, the Osprey was expected to require only a 35-hours inspection between major maintenance periods. It was hoped that a few years of operations could permit even this to be extended. Maintenance was to use Integrated Electronic Technical Manuals. These were essentially ruggedized laptop computers with interactive repair data ready for instant recall on the flightline or hangar deck.

The revised Osprey became the MV-22B and CV-22B, although the FSD machines were never officially designated A-models. External changes included projecting fairings at the aft end of the sponsons and on the lower quarters of the nose – the latter called 'chipmunk cheeks'. These oriented EW sensors and antennas while allowing them to be revised readily without altering aircraft structure. The cargo ramp was divided in two to create an upper cargo door and lower ramp, each with pairs of actuators. This would make 'fastrope' or rappelling operations off the ramp easier and reduce the cg shift with cargo rolling off the ramp during airdrop. It also improved the field of view (FOV) to the rear in flight for the flight engineer while checking aircraft condition, clearance from obstructions, and scanning for threats. Whereas the end of the original ramp was beveled to facilitate wheeled vehicles, the shortened ramp required ramp 'toes' to provide the smooth transition. The main landing gear were moved to the aft end of the gear well and turned 180° to help eliminate an aircraft tip-back potential during slope landings or rearward braking. The simple rectangular well main gear doors were segmented.

The interior saw new troop seats to accommodate the weight and bulk of a fully equipped soldier. More significant for the pilots was a displays redesign and modification of the TCL. The original controller had just 4in (10cm) of travel in a slight arc, making fine control difficult. The vertical grip, for some pilots, aggravated the tendency to mistakenly treat it like a helicopter collective. This played a role in aircraft 5's mishap with the pilot instinctively 'dumping collective' to 'plant' the aircraft on the ground when at one point its tires had met the pavement. Instead, the machine leapt back into the air. Another concern was that the lever's height made it likely to be inadvertently bumped when reaching for other controls or getting out of the seat with engines running. The redesigned TCL had a lower profile with a broad, horizontal grip and more comfortable palm rest. Fine control was easier, although it still had just 4 inches of travel. A 2-inch overtravel is available to compensate for power loss from a dual simultaneous mast torque sensor failure. Some of the analog standby gages were replaced with the digital Standby Flight Display. Changes were introduced in the cockpit layout and displays to ease operation.

System revisions addressed the Ice Protection System to improve performance plus reduce overall weight and cost. Avionics upgrades included a new mission computer that gave a 500% improvement in memory and speed. A 4% improvement in engine sfc was realized through changes in the turbine and compressor. The aft nacelle cooling inlets became scoops and mechanization to redirect IRS exhaust flow was replaced. The lighter system expelled air perpendicular to the exhaust flow to deflect the gases outboard. The development of self-deployment long-range tanks did not begin in earnest until 1999, although the fuel lines and valves in the cabin were incorporated from the beginning.

The initial EMD aircraft were flown without the heavy vibration suppression system, but it was quickly apparent that something would be required. The Active Vibration Suppression System (AVSS) had masses driven by electromechanical motors at frequencies determined by a computer using structural response measured in the cockpit. It sought to reduce or cancel vibration with an equal and opposite countering input.

New STA and FTA airframes were built to test the revised structure. Apart from Philly and Fort Worth, the work was also conducted in St Louis. The static article was subjected to over 200 individual drop tests and subsequently loaded to failure. Fatigue testing sought to demonstrate 2-lifetimes totaling 20,000 flight

Right: **The V-22 Integrated Test Team occupied the two hangars and attached offices seen in this image at Patuxent River NAS, Maryland, during EMD.** NAVAIR

hours for an airframe designed to 40,000 hours. The GTA was put to work again qualifying the revisions to the drive system for the transmission rating increased to 4,570shp (3,408kW). A proposed uprating of the engine to 6,450shp (4,810kW) was apparently found unaffordable.

Development and flight test was supported by software preparation and avionics integration efforts in several facilities. Greater use of simulation was also expected to make the testing more efficient. Philadelphia had three facilities supporting this goal. The Systems Integration Laboratory allowed actual aircraft avionics elements and software to be tested together prior to introduction on the aircraft. The Flight Control System Integration Rig used actual aircraft control system hardware driven by representative flight commands to validate design changes. The Flight Simulation Lab was a man-in-the-loop cockpit with flight hardware. These three labs could be linked together to evaluated the integrated system as thoroughly as possible on the ground. Engineering and training manned simulators included the V-22 Manned Flight Simulator (MFS) at Pax and the sims in Philly and Fort Worth. The V-22 Avionics Integration Laboratory was built at the Navy facility in Indianapolis. No full-up integrated systems 'ironbird' remained or was deemed necessary.

Flight test was to be concentrated at Pax to improve communication and reduce costs. The first test aircraft was scheduled to arrive in Maryland on 1 September 1996. Modern telecommunication tools were used to ensure fast and efficient information and data exchange within the far-flung engineering and management team. Approximately 350 personnel made up the Integrated Test Team (ITT). With the government working side-by-side with the contractors and their subcontractor, this was intended to reduce government-only testing. The government testers included Navy, Marine and Air Force personnel, both development and operational types. The 180-member Multi-service Operational Test Team (MOTT) included aircrew, maintainers and analysts.

Risk Reduction Testing

Following the aircraft 4 crash investigation, the Osprey did not take to the air again until 18 June 1993. Aircraft 2 and 3 were given safety enhancements before their next flights and further modifications to test EMD changes. No 3 flew again on 17 June and 2 the following month. They were devoted to EMD 'risk reduction' testing that was expected to see 135 hours accumulated on 2 and 200 on 3. This work concentrated on completing flight envelope expansion, airloads investigations, and the inflight icing trials. Collected data supported EMD redesign, allowed updating and validation of analytical models, and identified critical conditions to be addressed in EMD. It also allowed high-risk testing to be conducted with aircraft possessing ejection seats since none of the EMD articles were to be so equipped. A secondary objective was pilot training in preparation for the EMD. Although testing began in Arlington, the September 1993 decision to move all Osprey testing to Pax saw No 2 and 3 relocated to Maryland.

Aircraft 2 supported flight control system optimization, hover performance measurements, initial operational assessments, and demonstration flights. Also included were simulated aerial refueling tests. As the aircraft had only an instrumentation boom instead of a true AR probe (although of identical length), this work consisted of proximity flight behind a KC-130 hose and a KC-135 with a boom-to-drogue adapter. The trials fully expanded the AR envelope with the FSD flight control system and identified no handling qualities issues or detrimental interaction with the drogue or tanker. However, the 4.92-ft (1.50-m) boom was found to be too short for the left-seat pilot, at some seat heights, to see the probe tip without stretching and craning. The basket tended to disappear under the nose as the aircraft closed to the contact position. This increased the likelihood of a missed engagement, overrunning the drogue such that it got too close to the proprotors. If the pilot overshot the approach there were only about three seconds for the pilot to react and arrest the closure rate. This was considered too brief a span, especially since the nose prevented judgment of basket proximity to the proprotor arc.

The delayed inflight icing trials were performed in February and March 1994. These permitted ship 2 to experience maximum continuous icing conditions for ongoing development of the IPS. Special attention was paid to IPS performance at the engine inlets, rotors, the windscreen, pitot-static, and angle of attack (AOA) system. Evaluation of cockpit procedures for flight into IMC conditions was another aspect of clearing the aircraft for all-weather operations. The tests included high-speed flight behind a NKC-135A icing tanker fitted with a spray rig attached to the end of the flying boom that produced a cloud of water droplets at altitude where it would freeze and accrete on an aircraft following behind. Low-speed tests with a larger cloud were done in March and April 1994 in Duluth, Minnesota, behind the Army's CH-47 fitted with the Helicopter Icing Spray System rig. Operation of the IPS was verified along with the aircraft and engines' ability to tolerate icing. Required design changes

Left: **An EMD Osprey occupies one of the ITT hangars during late night maintenance. The nacelles and proprotors are in the 'maintenance' position, leveling the nacelles for ease of access.** NAVAIR

were identified and a further evaluation of the production IPS would be necessary at a future date at even colder temperatures to include flight into natural icing conditions where the entire airframe could accumulate ice.

Aircraft 3 was devoted to envelope expansion work to include airloads, aeroservoelasticity, high-AOA tests, initial height-velocity (H-V) characterization, rejected takeoffs, external loads, and sea trials. H-V defines the altitudes and airspeeds from which a safe landing can be executed given single or dual engine failures. Other trials saw external loads clearance, single-engine performance, initial autorotation tests, and the first off-field landings. Data on high-density altitude takeoff performance was collected with 2 during mountain flights in Hot Springs, Virginia, during August and September 1994. Attempts to recover a 3.8% hover performance shortfall included increasing the rotor speed to 103.8% and the transmission uprating for VTOL, addition of OLC, optimizing the IRS exit area, extending the maximum 'flap' angle to 75° from 67°, and eliminating 'aileron' action in hover. More than half the shortfall was regained.

The risk reduction testing had some distinct phases for clearly defined evaluations and documentation of system performance. These consisted of integrated testing (IT) in which the government and contractors worked together, and operational testing for dedicated government evaluations. The OT periods became milestones for LRIP decisions. The results were, of course, tempered with the understanding that the aircraft were not strictly representative of the eventual EMD machines.

Risk Reduction Flight Test Summary

Phase	Aircraft No.	Dates
IT-IIA	2 & 3	23 April 1993 - October 1994
OT-IIA	2 & 3	10 June - 7 July 1994
OT-IIB	3	28 September - 20 October 1995
IT-IIB	3	1 January 1996 - 31 March 1997
IT-IIC	3	1 September 1996 - 31 March 1997
OT-IIC	3	1 October 1996 - 30 May 1997

During IT-IIA, performed at Pax River, the envelope expansion testing was undertaken. This sought to clear the external structural loads, aeroelasticity, performance, flying qualities, and high-AOA aspects. Mission-essential systems were also characterized to include those associated with ship compatibility and initial IPS checks.

Bottom: **After years of labor the V-22 apron at Pax River was full of test aircraft at the end of the 1990s.** Author's collection

OT-IIA was conducted at Arlington, Pax, and Quantico. The two aircraft were flown for 14.8 hours during 21 flying days. Evaluations included confined area landings, simulated AR, formation flight, and night operations. The MOTT judged that the aircraft had the potential to become fully effective and suitable with continued development. Concerns were raised with cabin limitations that would restrict payload selection, high downwash velocities, and reliability shortfalls. Tactics development was begun and would continue in later tests.

The MOTT took possession of aircraft 3, by then the only flying V-22, for OT-IIB. This saw 10.4 hours and 8 sorties flown at Pax. The testers again found some operational suitability and effectiveness shortfalls, with concerns raised about performance, terminal area operations, and avionics. Downwash evaluations in various scenarios and overwater operations were also performed, with the effects of dust circulation around the aircraft evaluated. Overwater testing consisted of a build-down in altitude to 10ft (3m) at 20kts. This permitted an early suitability assessment for overwater SAR and helocasting. Helocasting, also known as 'soft duck', involves deploying swimmers and a rubber boat off the ramp while flying slowly just above the water. The build-down went no further because of sensitive instrumentation that could not get wet. Tactics development continued.

The early look at fastrope from the ramp was facilitated by installation of a shortened ramp representative of that planned for the EMD aircraft. A bar across the aft fuselage opening was installed to serve as a makeshift rope anchor. The troopers wore 100 lb (45kg) rucksacks. The testing revealed that a standard rope was deflected aft by the rotorwash, even with troops on it, such that the personnel could not come down beneath the aircraft as desired. Work began on a weighted rope that would be less affected by the downwash. Fastrope from the cabin door was also demonstrated. A rope ladder, another common troop helicopter device, was evaluated. The hydraulic personnel rescue hoist was temporarily installed in the forward door for a quick test. It was found unsuitable for a variety of reasons and returned for further development.

External loads testing demonstrated the ability to perform load hook-up underneath the hovering aircraft. Ballast of 2,000 and 4,000 lb (900 and 1,800kg) were carried on the forward hook only. VSTOL and CONV trials were performed, with the 4,000-lb load flown to 175kts at 30°-nacelle angle. Flight to 220kts with the external load set an unofficial world record. No difficulties were encountered during the limited evaluation.

Right: **A formation of EMD test aircraft provided a rare photo op. Both are fitted with instrumentation booms in place of their AR booms.** NAVAIR

Bottom: **Aircraft 9 sets down in a confined LZ during EMD trials.** NAVAIR

For IT-IIB, initial steady-state autorotation and power-off glide (engines idle) evaluations were performed. The 'autos' were performed at 95°, 45° and 0°-nacelle angles. While steep and at high sink rates, the results matched simulation. The requirement for a survivable emergency landing from a power-off glide or autorotation appeared to be met. Another phase of sea trials was also scheduled. Similar general work was performed during IT-IIC. At this point the flights shifted toward training new DT pilots and preparation for the next phase of operational testing.

OT-IIC saw 36.1 hours accumulated on 3 with much the same conclusions reached as previously evaluations. The results of OT-IIA and B, plus the preliminary results of OT-IIC, were a requirement for LRIP Lot 1 approval in April 1997. This drew some criticism since the sole remaining FSD aircraft was not representative of the B-model, and certainly not production-representative, and subject to many operating limitations.

Some refinements were incorporated in the aircraft and evaluated during the risk reduction work. Changes to the wing leading edge deicing boots were made to improve aerodynamic characteristics. Another important effort was resolving tail buffet associated with high AOA and yaw attitudes. Vortices generated at the overwing fairing were impinging on the tail. Analysis and wind tunnel results yielded strakes placed ahead of the wing leading edge roots. These redirected the vortices and reduced their strength. They also allowed removal of structural 'tuning' weights in the vertical stabs for weight-savings. Flight test on aircraft 3 in the fall of 1994 determined the best size and orientation of the strakes. Changes to the flight control system, to include a rotor load reduction feature, were also wrung-out on 3.

One other airframe remaining from FSD also found use. The fuselage of aircraft 6 was used for a December 1993 egress demonstration. This showed that fully laden combat troops and the aircrew could safely perform an emergency egress from the various exit points. The airframe was then used to mockup new aspects of the EMD design. It was subsequently parted-out and the hulk became a ballistics live fire test article, with its structure shot-up during ground trials with rounds of various types to evaluate survivability. Aircraft 2 was placed in flyable storage at Pax in October 1994. Its wing was eventually removed and shipped to China Lake for more live-fire testing. Its fuselage was used for the installation trials of new troop seats and mockups of the cargo deck fuel tanks. It, too, was finally consigned to live fire testing. Ship 3's last flight was a ferry to the American Helicopter Museum, West Chester, Pennsylvania, for permanent static display. Marking the end of the Risk Reduction phase, this came on 27 March 1997, within weeks of aircraft 7 arriving at Pax from Arlington. Aircraft 1, long grounded, was preserved at MCAS New River, Jacksonville, North Carolina.

Risk reduction added 343 hours on the FSD Ospreys for a total of over 1,100 in over 1,000 flights before the EMD aircraft took over. The aircraft had been to 252kts and 18,500ft (5,640m) in level flight, 322kts in a dive and to 21,500-ft (7,555-m) altitude, pulled 3.2G, and carried a sling load to 220kts. The team had demonstrated cruise airspeed at maximum continuous power of 265kts versus the requirement of 240kts. Some 85% of the flight envelope was explored and the team felt confident they understood the greatest technical hurdles remaining. The work allowed EMD flight testing to move ahead smartly.

MV-22B Flight Test

Assembly of the first MV-22B aircraft, number 7, began in Philadelphia during April 1995. The first major assembly, the wing torque box, was completed in Texas on 10 May. The fuselage was flown from the Philadelphia International Airport to Carswell AFB in a USAF transport to arrive on 4 December 1995 and then trucked to Arlington where the wing was attached within hours. Efforts to keep the program costs down played a role in the government funding just four flight test aircraft versus the six in FSD. The flight test was also planned for a markedly shorted two years. There was a distinct feeling in some quarters that the program was being unreasonably limited and rushed, again placing it at risk.

Testing of the MV-22B did not simply pick up where FSD and risk reduction trials left off. The numerous changes to the aircraft required nearly the full spectrum of developmental flight tests and system optimization.

EMD Test Aircraft

Aircraft No.	Principal Purpose
7	Flight loads, vibration, TF/TA radar
8	Propulsion and systems, flight controls, high angle of attack
9	Operational evaluation (USMC roles), modification for USAF variant evaluation
10	Avionics, operational evaluations, aerial refueling, sea trials

Each new EMD aircraft was flown to Pax following about 20 hours of shakedown in Arlington. Development and manufacturing delays slipped 7's maiden flight from December 1996 until finally accomplished on 5 February 1997, with delivery to Pax on 15 May. The first flight and delivery of each subsequent aircraft was also late, culminating with 10's ferry on 29 January 1998. As ferry of these aircraft to Pax was criteria for LRIP lot approvals, there was pressure to get them out of Arlington as quickly as possible. Consequently, all of the machines required many weeks of additional work at Pax before they could enter productive flight test.

Test planning was 'success oriented' with little or no margin for unexpected results or delays. Late start-up and numerous other problems cost the testers months of lost time. The program was once again under pressure. It had to show good performance to overcome the damaged image credited to it from FSD. So, the dates for the start of OPEVAL, deliveries to the first squadron, and other milestones were held to aggressively.

The program experienced the typical development surprises of varying severity. Most were corrected quickly or deferred to allow testing to continue provided there was no safety impact. Some of the design problems did precipitate groundings and delays, requiring minor changes to the aircraft and regression testing. The discovery of loose conversion spindles in March 1998 was probably the most troubling. This is the structural member upon which the nacelle pivots. Among the more mundane but significant findings were problems with panel clamps and fasteners impacting overall aircraft reliability. On the whole the aircraft were much more sound and operated better than the FSD birds. While reliability improved, it was still a burden on test productivity. The frequent maintenance periods and test equipment reconfigurations were contribu-

Top: **Lifting the new Light Weight 155mm howitzer (LW155mm, a 9,320-lb load) during a test flight in May 1999. It proved to be a very stable sling load under the V-22.** NAVAIR

Left: **Hover performance trials in Arizona during 1998 saw the aircraft hovered out of ground effect while tethered to the apron. Lift force was measured via sensors at the cargo hook.** Paul Shank

tors to schedule slips. Other impediments were delays in avionics software delivery, slow development of IPS hardware and software, and parts shortages as LRIP began.

Testing throughout 1997 saw a monthly average of only 7.1 flight hours per aircraft compared with the target 15 hours. A recovery plan introduced in 1998 brought an average 19.6 hours per airframe. However, so much essential testing remained at the scheduled end of EMD flight testing that the program had to be extended nearly a year.

The ITT goal of eliminating government-only testing to reduce program duration and cost meant there was essentially only one integrated test period, IT-IID. However, a separate operational test period in the midst of this work, OT-IID, required an interim evaluation of readiness for transition of the aircraft to OT testers. This period, IT/DT-IID, tested some of the more operationally oriented systems and capabilities to clear them for OT-IID. The IT-IID development testing continued while OPEVAL (OT-IIE) was ongoing. However, even this arrangement proved somewhat disruptive as the flow of testing was interrupted to clear capabilities for near-term evaluation or aircraft were occasionally set down for weeks to prepare systems and instrumentation for looming trials.

Significant flight control system changes had been introduced with the EMD aircraft such that nearly a full spectrum of stability and control and flying qualities tests were needed. The typical fine-tuning for optimal handling qualities in all phases of flight was undertaken. This included seeking effortless conversions and reconversions without adjustments in cyclic or TCL due to trim, pitch attitude, or altitude changes. Likewise, it was desirable that the 'beeping' between 100% and 84% rotor speed (Nr) for changes in flight mode, and autoflap surfaces retraction, be free of trim change requiring pilot compensation. The Opposed Lateral Cyclic was again evaluated. It was judged to be too great a penalty for the measured benefits and deleted. Lateral Swashplate Gearing was retained. For slope landing trials, larger and steeper concrete pads, at 6°, 9°, and 12° were built on sloping ground at Pax. Apart from handling qualities, the tests also allowed data collection of gear loads during sink rates up to 12fps (4mps).

The mission-representative evaluations lent interesting insight into capabilities. The basic sling loads envelope was cleared with little difficulty. The only surprises were an inadvertent load jettison and another instant when a twisted load began to oscillate and the crew chief chose to jettison. In 1998 the AR envelope began to be cleared with 'dry plugs' on the KC-130 tanker. The Special Insertion and Extraction rig was also tested. This involved flying the aircraft with up to four persons attached to a cable dangling from the aft cargo hook. Further helocasting trials were performed over the Chesapeake Bay. Parachute jumps and fastrope were conducted from the ramp.

Rotor performance trials took the aircraft to Hot Springs, West Virginia, in early 1998. High density altitude testing was performed with aircraft 8 at Fort Huachuca, Arizona, in September-October 1998. This included hover performance tests with the aircraft tethered to the ground to measure lift force. Critical azimuth testing in Arizona involved flying in translation along the runway at various headings while evaluating handling qualities. This saw the Osprey flying sideways and backwards at up to 45kts. It allowed the known PU/SS handling difficulties to be further quantified. Acoustic levels were measured by flying over a microphone array on the ground.

In early 2000 aircraft 10 was at CFB Shearwater, Halifax, Nova Scotia, for initial natural in-flight icing tests. Future improvements in the IPS meant that further testing would be required.

By 23 August 1998 the EMD testing had accumulated 316 flights for 627.7 hours. During

Above: **Slope landing tests employed specially prepared landing pads at different angles laid in at Pax River. The pilot has just set down level on a cross-slope preparatory to easing onto the slope.** Author's collection

Below: **The Special Patrol Insertion and Extraction rig allows up to four persons to be attached to a cable dangling from the aft cargo hook so that they can employ their weapons. Aircraft 10 tests the rig on the V-22 for the first time.** NAVAIR

MV-22 EMD Flight Test Summary

Phase	Aircraft No.	Dates
IT-IID	7 through 10	January 1997 - 30 April 2000
IT/DT-IID	7 through 10	24 August 1998 - 30 September 1999
OT-IID	9 & 10	1 September 1998 - 31 October 1998 and January 1999 - February 1999
OT-IIE (OPEVAL, Parts I & II)	11 through 14	1 October 1999 - 22 July 2000
DT	8, 10, 21 through 23	29 May 2002 - 2005
OT-IIF (pre-OPEVAL Phase II)	21 through 23 ?	April - June 2004
OPEVAL Phase II	uncertain	November 2004 - April 2005

January 1999 the EMD fleet passed the 1,000-flight hour milestone (2,200 total for the V-22). Basic envelope expansion was considered complete in July 1998. The aircraft had been taken to 342kts, 3.9G, 60,500 lb (27,442kg) GW, 25,000ft (7,620m) altitude, and flown with a 10,000 lb (4,536kg) sling load to 230kts.

The vulnerability of the Osprey's electronics to lightning strike was tested in June 1999 when aircraft 10 was placed in a shielded hangar at Pax, surrounded with a copper wire grid, and subjected to 10,000 amp charges. Electromagnetic compatibility/electromagnetic interference testing in a simulated shipboard EM environment was also performed at Pax with 10. This was necessary to clear the aircraft for sea trials.

The planned 30-day shipboard tests were performed aboard the USS *Saipan* (LHA-2) operating off Norfolk, Virginia. The landing trials began on 14 January 1999. It was planned that aircraft 10 would be operated from six of the vessel's nine helicopter spots under various crosswind conditions to examine dynamic interference issues of the complex airflow over the deck on the ability to precisely hover and maneuver. Flight tests would also look at sling loads, SEO run-on landings, STO, and night operations. An assessment of general movement of the aircraft on the ship, including the elevators and hangar deck, was performed.

The usual shipborne launch and recovery procedures were utilized for the MV-22 with some adaptation. As before, no deleterious deck edge effects were noted. A demonstration of the ability to move the aircraft and secure it to a 'slash' parking spot within 5 minutes of landing was an important aspect of the shipboard work, requiring BFWS. The ability to accurately taxi around the deck rather than always being towed proved a decided advantage.

Short take-offs were performed with great success, offering a means to get off with heavy loads when deck space permitted. The roll began ahead of the island and 100-200ft (30-60m) aft of the bow. At 70° nacelle, gross weights of 47,300 lb (21,455kg) and moderate headwinds, the aircraft was aloft with a ground roll less than one full turn of the main gear tires. Full power was applied promptly and the Osprey climbed away smartly. This work suggested that rapid operations could be developed in which the MV-22 landed on a stern spot, taxied ahead of the island, took on another load, and 'STOed off' while keeping the helo spots clear.

Observed was a tendency, when on the deck with rotors turning, to develop a roll in rotorwash from helicopters. As an interim measure the two spots ahead of the MV-22 were to be left vacant until this issue was resolved. Jerkiness in the lateral axis and horizontal darting became evident during precision landings with quartering headwinds above 20kts. This made the hover and landing task more difficult, occasionally requiring several landing attempts. This was not atypical for rotorcraft in windy conditions and was considered tolerable. All went well until a left-seat landing on spot 7, forward of the port side elevator and opposite the aft end of the island. The aircraft began to drift slightly toward the island and the pilot put in left cyclic to correct. The V-22 rolled too far left, lagging the centering of the stick, and the pilot made a full right input to bring the wing up and full TCL to fly away. The aircraft had rolled to 37° bank angle just 10ft (3.1m) above the deck. Only the fact that the left nacelle was over the deck edge prevented a proprotor or nacelle strike. In examining the event it was learned

Above: **Dual fastrope from aircraft 10 shows how the ropes can be blown a bit aft, but this is not a severe impediment nor atypical of large rotorcraft.** NAVAIR

Left: **The first jumps from the V-22 were performed by the 2nd Recon Battalion, II Marine Expeditionary Force, from Camp Lejeune, North Carolina. They executed 24 free-fall jumps off the ramp at 10,000ft and 120kts during several sorties.** DoD

that the high lateral workload in the crosswind conditions, with flow off the superstructure drawn into the right rotor, had exceeded the control system's capability, aggravated by inadequate lateral trim authority.

Months of control law adjustment and shore-based evaluations, both in simulation and flight, resolved the instability seen aboard *Saipan*. There was pressure on the team to move quickly as successful completion of sea trials was a criteria for approval of the Lot 3 LRIP and long-lead funding of Lot 4. Tests to measure improvements in aircraft responsiveness focused on a lateral repositioning task in hover. An attempt was also made to examine once again the long-argued deck edge effects while hovering above stacked shipping containers under various wind conditions. Some change in lateral trim was required in this asymmetric condition, but it left plenty of control margin.

Adequacy of flight control changes was successfully demonstrated during a return of 10 to *Saipan* in August 1999 off Norfolk. LSD operations were conducted in September onboard USS *Tortuga* (LSD-49) with its two helo spots. The 75 launches and recoveries completed the embarked sea trials. The aircraft had performed 642 day and night shipboard launches and recoveries for 98 flight hours.

Sea trials cleared basic launch and recover envelopes for OPEVAL. It did not, however, clear the aircraft for takeoff or landings on the LHA's spots 5 and 6 directly opposite the island. Unanticipated protuberances on the structure were found within the clearance zone and the decision was made to postpone this testing. This would probably never be done on the LSDs because the vessels were scheduled for retirement within a few years. However, this requirement had largely sized the aircraft and rotor system. Lower disk loading and improved performance may have been possible without this constraint.

Six MOTT pilots participated in OT-IID at Pax, New River, and Eglin with 142.6 hours in 63 flights on ships 9 and 10. This was exit criteria for approval of LRIP Lot 3 and contract implementation for Lot 4. For this purpose Critical Operational Issues were assessed, including reliability, maintainability and availability (RM&A), sortie generation, and logistics supportability. There were still system limitations and immaturities when 9 and 10 were handed to the MOTT, but valuable insights were still obtained.

The four MV-22Bs and risk reduction A-models logged about 3,600 flight hours by the time of OPEVAL. Each B-model had flown about 400 flight hours for 711 flights totaling 1,469.6 flight hours. This was markedly fewer hours than most new military aircraft development programs. By mid-2000 aircraft 10 had been grounded to serve as a maintenance trainer. The four Lot 1 LRIP articles were delivered in 1999 in preparation for operational testing.

Top: **External loads testing began with simple ballast configurations such as this girder on a dual sling. The hook doors appear to have been removed.** NAVAIR

Right: **In conversion mode with partial flaps, aircraft 8 carried a 6,300-lb 'Humvee' across the Chesapeake Bay near Patuxent River NAS at up to 130kts. On one flight the windshield of the vehicle shattered because of the high speed.** NAVAIR

Above: **Flying low over the Chesapeake Bay near Pax, ship 10 raises a 'rooster tail' of spray.**
Chris Seymour collection

Readiness for OPEVAL was based upon IT/DT-IID assessment. EMD was declared complete on 30 April 2000. This, however, left many planned tests incomplete. It was judged that sufficient understanding of the aircraft and a suitable envelope had been cleared for OPEVAL. The MV-22B still failed to meet some maintainability and reliability requirements and was still essentially a developmental vehicle. This is not atypical of a new system, especially one in which all the BIT functions were not working properly. A number of significant capabilities were not cleared. Some were to be approved based on further EMD testing before the end of OPEVAL so that they could be included in the evaluation, while others would take longer. All this would be considered in the MOTT's report. This is not unusual and the Navy claimed that the 22 deficiencies/waivers were actually fewer than any other aircraft in their history.

Among the outstanding issues was a disconcerting tendency of the aircraft to momentarily settle when interim power was deselected. Interim power increased mast torque to 109% at 104% Nr for heavy weight or hot day operations. NBC protection needed further work, the cargo handling system was incomplete, airdrop was prohibited, and the rescue hoist was again found unusable. Efforts were still underway to reduce noise and vibration levels, and BIT had a high false alarm rate. Flight into icing conditions and air combat maneuvering were still prohibited. Additional operational limitations were issued during OPEVAL based on ongoing flight test.

Additional icing and AR trials were to be performed concurrent with OPEVAL, along with other ongoing clearance activities. Attempts in October 1999 to clear the aircraft to take gas from the Air Force's KC-10 proved fruitless. Refueling from the tanker's wing pods felt uncomfortable to the pilots because of the KC-10's wingtip vortices interacting with the outboard proprotor. Refueling from the center drogue could not be safely accomplished because of KC-10 AR system difficulties at the low airspeeds. A few requirements were also changed prior to OPEVAL, including increasing the land-based STO from a required 500ft (150m) to 3,000ft (915m) – somewhat challenging the definition of 'short'.

OPEVAL, originally planned to begin in January 1999, commenced in November and was to run through June 2000. The MOTT operated under the auspices of HMX-1 at Quantico whose mission included USMC rotorcraft operational test. They were to use the first four LRIP machines with a planned 700 flight hours during 350 missions at Pax, New River, China Lake, MCAS Yuma, Arizona, Hurlburt Field AFB, Florida, and other sites. A single aircraft was also sent to Kirtland AFB, New Mexico, for evaluation by the USAF's 58th Special Operations Wing. OPEVAL included the full scope of permitted USMC operations to include week-long exercises with all four aircraft aboard the USS *Essex* (LHD-2) off the West Coast of the US. Another round of trials was performed off the East Coast. At Yuma, the work included the transport of over 700 troops in 40 airlift missions. The OPEVAL report was expected in October to support the FRP decision that month.

The program suffered a tragic setback when aircraft 14, the newest MV-22B with just 135 flight hours, crashed on the night of 8 April 2000 outside Tucson, Arizona, during an OPEVAL exercise. The 19 Marines aboard were killed when the aircraft rolled over and dived into the ground from 245ft (75m) while making a landing approach to the Marana Northwest Regional Airport. The lead ship (11) in the formation of two, about 1,000ft (305m) ahead, landed hard (about 15fps/4mps), rolling off the paved surface and through a shallow ditch that tore off the AR probe. The V-22 test fleet had flown some 3,000 hours since the last crash eight years before, had been to 342kts, 25,000ft (7,620m), taken off at 60,500lb (27,440kg) and seen 3.9G.

An examination of 14's wreckage revealed no mechanical faults. The investigators concluded that the formation had arrived at their destination late and 2,000ft (610m) high. The lead pilot descended at a high rate to land on the first approach rather than going around and increasing tardiness. The effort at executing a steep descent with a tail wind caused the pilot to reduce forward speed, robbing the aircraft of translation rotor lift. This produced vortex ring state (VRS).

VRS occurs when a rotorcraft descends at a high rate with low forward velocity. The rotor is descending through its own downwash with the blades experiencing a very high inflow incidence angle. Adding power (increasing blade angle through collective) only exacerbates the problem until the blades stall. Alternatively, when the descent rate approaches the downwash velocity, the air is moving up through the rotors as fast as or faster than it is being pushed down. The rotor loses lift. The descent rate increases and cannot be arrested even with full power. Because of blade twist, the inner portion of the disk stalls first. This produces an upward column of air while the outer portion of the disk continues to produce downwash. This opposing motion results in a recirculating vortex of air through the disk in a ring pattern. The common recovery procedure is to reduce power and push over with cyclic to gain forward velocity and translational lift, flying out of the disturbed air.

Aircraft 14's steep descent was not in conformance with the flight manual limit of 800fpm (4mps) below 500ft (152m) at less than 40kts (an adopted CH-46 operational limit) – although the presentation in the manual and the consequences were not well understood. The pilot, seeking to maintain his position on the lead, was descending at 1,800fpm (9mps) with the nacelle back at 95° and the nose up at 5-7°. In addition, the 8-15-kts tailwind had as much as a 5-10-kts gust component. These factors produced an extremely high proprotor incidence with respect to the inflow. This was increased even further on the right rotor when the pilot corrected his heading with a 5-15° right banking turn with a bit of coordinating right pedal. The right rotor lost lift first and the aircraft dramatically rolled and yawed right. The low altitude did not permit a safe recovery, even if the pilot had recognized the VRS condition. Aircraft 11 apparently also suffered a loss of lift, causing its hard landing.

It is possible that 14, flying trail, also encountered disturbed air if it descended below the lead. This is a recognized rotorcraft phenomenon that can produce a strong downward force. On two other occasions during OPEVAL pilots reported uncommanded rolls during formation flight. Helicopter formations commonly land lead-first to ensure that trailing aircraft do not fly through the wake of those in front.

Above 70° conversion angle the V-22 proprotor is subject to stall the same as a helicopter rotor. Although never intentionally investigated on the XV-15 or V-22, the highly loaded rotor with considerable blade twist, and limited test results, had led some to believe that VRS was not a tiltrotor characteristic. Others claimed that the large blade twist exacerbates VRS potential. The lateral tandem rotor layout will mean sharp rolls and thrust fluctuations should one rotor enter VRS before the other due to increased cyclic or yaw. As the aircraft rolls, the down-going rotor experiences a further increase in airflow velocity and angle through the rotor. The pilot will naturally apply opposite cyclic to raise the wing, further exacerbating the effect. The aircraft rolls over uncontrollably. Consequently, tiltrotor VRS is inherently more critical than a helicopter, yet recovery can be more positive simply by rolling the nacelles forward as little as 15°.

Investigation of the accident led to the 800-fpm descent rate limitation being applied whenever the nacelles were aft of 80°, regardless of airspeed. Pilots were also instructed not to fly closer than 200ft (61m) behind another Osprey nor 50ft (15m) below (another CH-46 restriction). These prohibitions raised concern that they would prevent the operators from realizing the full potential of the aircraft. However, it was hoped that subsequent testing would ease the restrictions.

The investigation team took NAVAIR and DoD leadership to task for conducting only a small portion of their high rate of descent (HROD) testing. Planned test points were halved to accelerate EMD, but then only two thirds of these were actually flown – those essential for showing the adequacy of the placard. Anything more was considered unnecessary given schedule pressures. Although such adjustments are practical based on analysis of collected data, such cuts had apparently become endemic of the V-22. With so little understood about HROD effects on tiltrotors, they appeared unwise in hindsight.

Getting the aircraft into OPEVAL and meeting IOC on time had become paramount. Emphasis on completing a thorough engineering test program preparatory to investing billions of dollars in production had unwittingly become secondary. It subsequently became an open question, echoed in a GAO report, whether reduced testing had sufficiently documented the performance, characteristics, and safety of the system prior to the machines being handed to the MOTT. At the same time, criticism was leveled at the operational test community for too readily accepting test articles with incomplete development testing and significant limitations. Even some development testers felt it was more than likely the MOTT would suffer an incident because of what they perceived to be precipitous transition to OPEVAL.

The crash raised another hue and cry from those who felt the Osprey fatally flawed. Heard again were the familiar criticisms that the aircraft was too complex, too radical, too hard to fly, too expensive, and simply the wrong choice for the Marines. The investigation results and steadfast adherence to the Osprey by the Marines allowed the din to fade away without risk to the program. The V-22 was becoming known as the 'Teflon Weapon' because none of the criticisms stuck. Still, it appeared to some, with three V-22 crashes prior to deployment, that the aircraft might experience a similar high number of losses and deaths in USMC service as did the Harrier before it was learned how to operate that machine safely.

The MV-22s were flying again on 19 May 2002. Additional testing was undertaken to characterize VRS potential and the optimal recovery technique, in addition to wake influence in formation. Testing for the latter included mapping the wake by flying over an array of laser beams with the aircraft at various conditions and configurations. During HROD work the conditions evident in the crash were reproduced and resulted in a dramatic departure from controlled flight. The testing and VRS studies suggested the V-22 was not especially susceptible to VRS and could be safely flown to more than 1,400fpm (7mps) at 30kts. Although there appeared to be more that could be learned about tiltrotor rotor stalls, enough was revealed to safely operate the V-22.

Aircraft 15 was assigned to OPEVAL when it was delivered in June, replacing 14. Resuming on 5 June, OT-IIE ended on 22 July after 522 sorties and 805 flight hours, and carrying more than 700 troops. The OPEVAL report was released on 13 October 2000.

The OT team found inadequate situational awareness for the cabin crew because of limited FOV, poor lighting under certain conditions, some lack of suitable warning, caution, and advisory annunciations, and other factors. Although the cockpit was kept comfortable, it was difficult to cool the cabin to desirable temperatures. On the positive side, the V-22's speed, range, and handling qualities far exceeded medium lift helicopters. All key performance parameters were met and, in many cases, exceeded the program threshold requirements. Maximum cruise speed was measured at 258kts versus the required 240. The minimum amphibious external lift requirement of 30nm was exceeded by 20nm. On a typical Marine mission the aircraft flew its 50-Nm distance with 11,700 lb (5,307kg) of payload compared with the objective 10,000 lb (4,536kg). The STO distance on a dry and hard runway was expected to be 950ft (depending on GW) compared with the requirement of 3,000ft (914m), and 140ft (43m) versus 300ft (91m) shipboard. The aircraft would complete a self-deployment 4 hours faster than required. It was anticipated the 2,100nm self-deployment would be exceeded by 161-179nm. The mean turn-around time requirement of 15 minutes was actually just 8 minutes.

Right: **Flying very low and slow over the water to deposit the 'soft duck' swimmers and boat, the Osprey raises a cloud of recirculated spray.** Author's collection

Although deficiencies were noted and some tasks could not be performed because of waivers, the aircraft was found operationally suitable and effective for the USMC mission. Of 243 operational requirements, the aircraft had failed to meet only 17. The remaining issues with the aircraft were considered minor and efforts were already underway to address them. However, the testers concluded that the MV-22B was not suitable for sea deployment because of maintenance and serviceability problems. The principal complaint was unreliable BFWS. Frequent failures affected maintenance accessibility and the ability to move the aircraft below deck. The most significant reliability issues centered on the hydraulic and the drive systems.

Of the RM&A metrics, only 2 of 12 were unequivocally met. Most significant was that the machine fell considerably short of the threshold requirement of 82% or more Mission Capable Rate, seeing only 57%. Consequently, the conclusion that the Osprey was suitable seemed so discordant to some in Congress that they were calling for an investigation. Proponents countered that the aircraft had met the more significant 17-hours Mean Flight Hours Between Aborts. However, the MV-22 was simply not as reliable, and required more labor to maintain, than the helicopters it was to replace. But the V-22 RM&A requirements were among the most demanding levied on any aircraft program. One of the waivers going into OPEVAL was for reliability, it being clear that the aircraft would fall short. The program office planned to use the reliability data to target improvement efforts. Most major weapon systems go through a cycle of improvements over the first years of service to reach desired reliability and readiness levels.

Despite the controversy, the OPEVAL conclusions were allowed to stand.

It was critical that BFWS reliability be improved and demonstrated quickly so that vital shipboard suitability would be clear before the Pentagon formally considered full-rate production. Bell Boeing worked diligently on fixes. An MV-22 deployed to the USS *Bataan* (LHD-5) on 31 October 2000 for a day of BFWS evaluations that it easily passed.

The program was hoping to make their case for FRP in early December 2000. However, the DoD's director of Operational Test and Evalua-

Top: **Aerial refueling trials against the KC-130 went smoothly, and the capability was cleared for use during OPEVAL.** NAVAIR

Middle: **Aircraft 8 performed proximity tests against the KC-10 in late 1999 prior to actual refueling. The Osprey is approaching the basket as seen from the boom operator station.** NAVAIR

Left: **Attempts to clear the V-22 for AR contacts against the KC-10 tanker were frustrated by drogue stability problems, but the Osprey (10 shown) handled well at the centerline position of the jumbo jet.** NAVAIR

Above: **Countermeasures dispensing tests verified safe separation from the aircraft when firing flares and chafe from the aft sponson buckets.** NAVAIR

Right: **This extraordinary picture of a folded Osprey on an elevator affords an excellent view of the wing vortex generators, wing cove panels, mid-wing accessories hatch, and APU exhaust.** NAVAIR

tion, Phil Coyle, reported he could not support the decision. In his opinion the low RM&A results and other issues made the aircraft operationally unsuitable. He also felt enough had yet to be done to address VRS. A separate DoD Inspector General report echoed some of these concerns. The Navy responded by pointing to the more than 118 system changes already in work to address these problems and others. The USMC leadership indicated that, since the Osprey was new, the experience level in maintaining the aircraft was still comparatively low and this contributed to the poor showing.

At this point the program was again thrown into crisis when the newest MV-22B, ship 18 with just 157.7 hours, crashed on 11 December. Four Marines from the new training squadron, Marine Medium Tiltrotor Training Squadron VMMT-204 (previously HMT-204), MCAS New River, were killed. With about 880 hours logged since the last accident, the Osprey was once again grounded.

With the press aggressive in highlighting Osprey faults, the USMC deferred the production decision indefinitely. The service called for an independent investigation of the whole program while the Navy began the mishap and legal culpability investigations. The second fatal crash in 2000 raised many old questions anew. The public and even the military had become used to advanced aircraft being developed and deployed without accidents.

With the unique tiltrotor V-22 incorporating so much new technology, especially as a production article, such expectations were perhaps unwarranted.

The program suffered another blow in January 2001 when the Marine officer commanding VMMT-204 was relieved of duty. The action followed an anonymous letter posted to the Secretary of the Navy by a mechanic in the unit. It accused the Colonel of requesting his personnel falsify records to conceal low aircraft readiness and high maintenance manhours. The assertion was supported with a tape recording of a meeting in which the officer openly admitted they had to lie to save the program.

The DoD Inspector General took responsibility for the investigation into the readiness 'cover-up'. In addition, the GAO launched its own investigation into the wisdom of full-rate production. This brought to five the number of concurrent investigations of the Osprey underway in early 2001. Hearings on Capitol Hill were certain to follow. The aircraft was unlikely to be cleared to resume flight operations until all reports were in and recommendations had been considered.

Aircraft 18 went down at night, in VSTOL, five miles short of the New River runway. The close routing of tubes, hoses and wire bundles within the tight confines of the nacelle, combined with high vibration levels, contributed to chaffing of a hydraulic system No 1 line in the left nacelle until the titanium tube burst. This line provided pressure to the three swashplate actuators in the nacelle. But, losing one of three hydraulic systems should not have rendered the aircraft unflyable.

For redundancy, each swashplate actuator is powered by No 1 and 2 of the three hydraulic systems. The electronic controls detect rapid fluid loss and takes action to isolate No 1 pressure from the damaged portion of the system. The No 3 utility system then takes up the load to maintain pressure to the actuators, but with a reduced actuation rate. Consequently, the left hand swashplate responded at a lower rate than the right proprotor swashplate during the mishap.

The hydraulic system switchover and difference in swashplate rates activated the primary flight control system PFCS FAIL/RESET switch/light and tone in the cockpit. The pilot, following procedures, pressed the switch to reset the PFCS. Because of a vehicle management system software logic error, this momentarily caused

Top: **Aircraft 10 rises to the flight deck on one of the USS *Saipan*'s elevators. Note the vertical stabilizer trailing edge light, formation light, top anti-collision beacon, and rudder balance weight horn.** NAVAIR

Left: **The V-22 projects beyond the edge of the LHA's stern elevator as a crew prepare to tow it onto the flight deck. The Osprey's nose strut is 'hiked' up to allow the tow bar to be installed.** NAVAIR

the proprotor governor to move all six blades to zero pitch. The aircraft lost thrust and decelerated. The system recovered quickly, but the return of the proprotors to their normal state caused a sudden aircraft acceleration. The left proprotor, operating partially on the utility system, recovered more slowly than the right. This saw as much as 10° difference in collective and a violent yaw to starboard. The PFCS FAIL/RESET reoccurred and pilot reset eight times. This repeatedly caused the aircraft to decelerate and accelerate, accompanied by strong yaw, roll and pitch. The longitudinal forces also produced involuntary pilot inputs to the TCL, compounding control difficulties. The aircraft eventually stalled and descended into the terrain. While more than three attempts to reset the system was contrary to good practice, the crew had inadequate insight into what was happening and little chance of recovering to a safe landing in any event.

The recommendations of the accident board ran from correcting the software fault to redesigning the hydraulic system. The latter included changes in the nacelle to reduce chaffing, a substitution of material for the local hydraulic lines, and redesign of the lines to the swashplate actuators to enhance triple-redundancy. They criticized testing for not having examined the contributing failure conditions and took the designers to task for having inadequate understanding of the effect of proprotor rate changes on controllability. Although a Crash Survivable Memory Unit (CSMU) recorded flight data during the accident, the investigators judged it inadequate.

The Navy promptly set to work on the software problem and nacelle changes. Efforts were already underway to replace the CSMU with a more capable unit that also included cockpit voice recording. Criticism was leveled at the 5,000-psi hydraulic pressure and titanium lines. These and the redundancy/failure features had undergone detailed safety analysis and many aspects were common in rotorcraft. Fundamental hydraulic system redesign and other proposed changes to address different swashplate actuator rates in a failure state, if deemed essential prior to FRP, would be costly and significantly delay the program. All would need to be carefully considered.

Much erroneous and misleading reporting in the press fanned a flurry of attacks on the V-22. Opponents pointed to four crashes with 30 deaths, 15 years of development and extensive testing, roughly $12 billion invested (the Pentagon's sixth largest development program), and the aircraft was still not ready for fleet deployment. All was held forth as evidence the program should be cancelled. The varied nature of the four accidents, none owing to the unique nature of the tiltrotor, was largely ignored. The on-again off-again nature of the Osprey's development was not reviewed. Few comparisons were made with other military rotorcraft development programs and early deployment to reveal that the V-22 was not especially

Top: **The Osprey should seldom be seen in this configuration aboard ship, with the nacelles lowered level to the 'maintenance' position. But, flight test brings with it all manner of unusual things.** Boeing

Right: **During sea trials aboard the USS *Saipan*, aircraft 10 performed all common shipboard tasks including external loads. Here aircraft 10 hovers with a 6,300-lb 'Humvee' under a dual sling.** NAVAIR

accident-prone. Few military warplanes were without things to be fixed at time of production and fielding, and it was unreasonable to expect the V-22 to be perfect – although certainly safe. Likewise, the readiness of the V-22 was not clearly compared with other warplanes at a similar stage in development. Such analysis would have shown that the Osprey was not in particularly poor shape.

Opponents again urged the often-heard suggestions of V-22 cancellation, reduction in production numbers, or substitution of existing helicopters. The cost figures offered for the alternatives did not account for the changes required to adapt them to the Marine mission. Even if the Osprey continued, voices called for at least another year improving reliability and additional testing before FRP and IOC.

The independent 'Blue Ribbon' panel reported on 19 April 2001 with a mix of good and bad news. It found the V-22 of value both to the military and the US economy. It filled requirements in a way that no alternative could. Starting over with another aircraft would take many years and hundreds of millions of dollars to get to where the V-22 was then in its development, and likely with a less capable machine. They concluded that the basic tiltrotor technology was not flawed. The manufacturers should be allowed to keep the production line 'warm' and their vendors committed. However, production was to proceed at the lowest practical rate so that few aircraft would later need to be modified to the final configuration. The recommended nacelle changes were primarily to simplify access for easier inspections, in addition to the hydraulics issues. The panel also sought further development and testing to improve reliability and maintainability. They suggested tests and training to deal with VRS and asym-

Top: **Tests to measure improvements in aircraft responsiveness focused on a lateral repositioning in hover. The aircraft was aggressively translated sideways and abruptly returned to a hover to precisely lineup the poles seen here.** NAVAIR

Left: **During the first OPEVAL the four MV-22Bs were given special markings that included the HMX-1 logo and a number of 01 through 04 appearing on the sponson forward face and tail. Ship 11 is shown.** Author's collection

metrical rotor lift. Associated was the suggestion for a cockpit warning system to help avoid the hazardous regime. The panel's report pointed to a major program restructuring.

A NASA panel formed to examine tiltrotor aeromechanics phenomenology from a technical perspective reported in November 2001 that they also found no 'show-stoppers' and echoed many of the recommendations of the Blue Ribbon Panel. They called for more testing and analysis of VRS, formation, and shipboard flowfield effects, but also suggested dropping the requirement for autorotation landing to a survivable landing under all conditions. Apart from being virtually impossible in the V-22, it was met by few if any multi-engine transport helicopters. The Marines had to deliver in December 2001 yet another analysis of alternatives to the MV-22. As anticipated, the GAO recommended additional testing and reliability improvements before commitment to production. Their preliminary report also accused the Navy of omitting or curtailing vital testing, casting into doubt the safety of the aircraft and readiness for FRP and fielding. The criticism of deferred or deleted testing was echoed by a Defense Science Board report released in February 2002. It pointed to the budget and schedule pressures, usually from outside the program, driving managers to cut corners in testing. Fallout from these observations was that a board of senior Pentagon personnel assumed responsibility for high-level program decisions.

Top: **Like an artist concept come to life, OPEVAL saw four Ospreys in settings only dreamed of just a few years before.** NAVAIR

Right: **Just rising from the bowels of the ship, aircraft 14 is ready to be towed onto the deck and prepared for another flight during OPEVAL. Note the covers over the nacelles, exit faces.** Ron Culp

The fate of the program rested with the new Bush administration. It was conducting a detailed review of military requirements during the many V-22 investigations. It was clear the Marines needed a new aircraft and starting another program would only add cost and time during a period high operations tempo was wearing out equipment at an accelerated rate. This was especially so following the 11 September 2001 terrorist attacks on New York City and the Pentagon. Operations in Afghanistan required the Sea Knights to fly long distances from their ships to the land-locked country. The USMC's 229 CH-46 helicopters were suffering dramatically rising operating costs. Likewise, the USAF had to adjust its force planning to retain the MH-53s a bit longer while awaiting the CV-22. However, the war on terrorism did not greatly reduce the threat to cut major defense programs, especially those that appeared designed for major battles with heavy equipment rather than 'transformational' weapons in the new era where highly mobile light forces appeared to be paramount. Pundits managed to cast the Osprey in both categories, with the SOF mission becoming more important in the new era.

The mishap opened many issues about the thoroughness of past failure mode analyses, ground tests, and adequacy of the flight test. The wisdom of not testing failure modes in flight was questioned. Analyzing the ship 18 event, other failure states, and their consequences required improvement in analytical tools and simulators for suitable accuracy and fidelity. The work especially focused on hydraulic failures and their affect on aircraft handling, changes to cockpit indications for situational awareness, and revised emergency procedures. The software and hardware revisions required many months of design and ground testing. No one felt comfortable operating the aircraft until this was done. Thousands of hours of evaluations provided valuable insight, yielding additional software and hardware changes or long-range plans for same.

Although the V-22 program expected to resume flying in April 2001, this proved optimistic. A healthy openness in examining all issues raised by program personnel and outside reviewers revealed much work to be done. Efforts were made to ensure the work was event- and not schedule-driven. Teams from the services and contractors examined all systems for latent design flaws. A list of hundreds of changes resulted, but careful assessment concluded that only a subset, termed Block A, had to be corrected, with favorable flight test results, prior any training aircraft returning to flight.

Block A was to included annunciation of flight beyond 800-fpm rate of descent and less than 40kts. It was hoped this limit would be extended, especially as it impeded certain CV-22 TF operations. The test aircraft would get the equivalent of the Block A and begin flying when all was ready. Others would follow, with VMMT-204 not expected to resume operations before December 2003. The remaining changes were broken into Block B, with enhanced suitability and nacelle maintenance access. C contained improved mission capabilities, provided funding was forthcoming. All were to be made as modifications to existing aircraft or introduced into the production line over the coming years. Block B would be introduced in 2004 and C in 2007. These efforts only added to the persisting weight problem and certain performance deficits, prompting reenergizing of weight-reduction efforts to meet the adjusted 33,140 lb empty weight specification.

Above: **A brand new MV-22B, aircraft 13, cruises in airplane mode a few hundred feet above a river estuary during OPEVAL in 2000.** Ron Culp

Flying only resumed after numerous high-level reviews and approvals. The modified and much-inspected aircraft 10 finally took to the air at Pax on 29 May 2002 after nearly 18 months grounded. Ship 8 following on 19 October with HROD its principal focus. Planning had settled on 18 months and about 1,800 hours of MV-22 testing to complete EMD in 2004. Five fleet-representative aircraft, 21-24 and 34, joined the Pax work as trainers and supplemental test aircraft. Aircraft 21, the first production machine modified with the Block A changes, flew on 7 September 2002 and ferried to Pax on 12 October. 34, the first produced with Block A, arrive on 20 August 2003. By 30 May 2003 the continued EMD testing had added 500 hours of V-22 flying – and another 500 by 5 December – for a total of about 6,400 on all V-22s.

The planning and flying in 2003 was very productive and without major incident, restoring the confidence of many. Testing wrung out design changes to enhance safety and reliability, continued HROD work, looked at formation influences, and reopen many incomplete test plans. The Pentagon also called for another

examination of combat maneuvering, AR, and dust and debris recirculation effects during landing. 10 went out to the USS *Iwo Jima* (LHD-7), off Norfolk, on 14 January 2003 for five days. This continued with 10 and 22 on an 11-day deployment to USS *Bataan* in November 2003. The team gathered more data on the uncommanded roll-on-deck phenomenon from rotorwash of other aircraft. Another phase of shipboard trails was planned for April 2004. The Navy had engaged industry and academia to model deck flowfield effects on the V-22 and understand the influences. Software modifications made the lateral damping feature of the flight controls active on-gear to see if this helped mitigate the effect. Additional work to reduce PU/SS was undertaken. Aircraft 10 would also perform additional AR testing, 10 and 22 work on mission systems tests, 21 on night formation, and 21 and 22 on austere landing trials. 21 was also used in developing the airdrop capabilities and looking at the associated cg shift, with a deployment to Fort Bragg, North Carolina, in January-February 2003. This included parachute-recovered loads of up to 2,000 lb (910kg) and paratroop drops. Aircraft 24 was to get a faux CV-22 radar for testing of radome deicing along with the complete natural icing test of the IPS in Nova Scotia during December 2003. This was to continue the following winter.

Another OPEVAL was to be proceeded by an OT-IIF assessment period. The MOTT was reestablished for OPEVAL Phase II planned for November 2004 through April 2005 in which RM&A would be vital measures. Its work would be under the auspices of a new unit, VMX-22, established at New River on 28 August 2003. It was anticipated VMX-22 would continue testing MV-22 upgrades for many years with 4 of its 16 aircraft. It would provide aircraft and personnel to VMMT-204.

Top: **Aircraft 21, enroute to Pax River during October 2002, prepares for departure from its refueling stop at Wright-Patterson AFB, Ohio. Note the lack of an AR probe.** Navy

Right: **Cargo airdrop tests from the MV-22, aircraft 21, were performed in early 2003. The extraction parachutes on the two bundles have inflated to pull out the recovery parachutes.** Navy

CV-22B Development and Testing

Planning for the CV-22 resumed in 1995 and the contractors' proposal was submitted that December. But, it was January 1997 before the deal was cut and half of that year spent formulating the EMD plan. Bell Boeing had sought $780 million for the effort but the Navy was able to commit only $490 million. The EMD aircraft had not been built to support CV-22 testing. However, savings would be realized by remanufacturing MV-22 to the CV standard after completing the bulk of MV testing.

The USAF was faced with an aircraft optimized for the Marine mission and so, as has become traditional for SPECOPS, had to adapt the machine. As always planned, the CV-22B would feature additional wing fuel cells extending flight time by 1.5 hours for a typical mission. The ability to install these cells was part of the baseline V-22 design, but the other features would require reengineering the aircraft. The USAF was also compelled to reduce its performance requirements for the realities of the established Osprey platform and AFSOC 'operators' would also have to adapt to the airplane. Markedly different performance and interior dimensions than the MH-53 would make some existing equipment and tactics unsuitable. Even the oxygen equipment and headsets were nonstandard.

The CV-22 would require a more capable EW system, the rescue hoist, plus enhanced navigation and communication systems. The flare/chaff dispensers were to be replaced with USAF units. The AN/APQ-186 multi-mode radar (MMR) was selected along with the AN/ALQ-211 Suite

of Radio Frequency Countermeasures (SIRFC). Under development by the US Army, SIRFC promised 'cutting edge' EW capabilities greatly enhancing threat awareness and survivability. The USAF chose to delete the MV-22B's AN/AVR-2A laser-warning receiver, reducing weight and drag to meet the range performance and because of intentions to install a more capable system later.

Needing to flight test the advanced EW system and perform low altitude TF trials, a move to Edwards AFB, California, was approved as a cost-savings measure. Most American EW range assets are located in the west, and the majority of TF development flight tests are best performed over barren terrain and mountains. The Air Force Flight Test Center (AFFTC) had considerable TF and EW expertise to contribute. Military maintenance offered the opportunity to begin collecting RM&A data while operational test participants gleaned early insight into military utility. Because the Marines were interested in some aspects of the CV-22B, the Corps provided a small cadre of personnel. CV-22B development flight test was expected to begin in the latter half of 1998 and the first production example was to be delivered in 2000. The USAF made clear at the time that they reserved the privilege of renaming the Osprey.

Aircraft 7 and 9 were selected for CV-22 remanufacture. Both were flown back to Arlington in the summer of 1999, months late owing to the press of urgent MV-22 testing preceding OPEVAL. Ship 7, to serve principally for TF/TA development, was given the auxiliary wing tanks and radar. Aircraft 9 was given the full complement of CV-22 system plus the AVSS. Replete in a new paint scheme, it was featured in the formal rollout of this first CV-22B on 25 July 2000. The CV-22 test began with the next flight of 7 on 29 February 2000. Following shakedown testing, it was flown to Edwards on 20 November. 9 flew again on 21 August and ferried on 18 September.

The CV-22 testing evolved into one integrated test period (IT-IIE) and Initial Operational Test and Evaluation (IOT&E) with a planned 1,182 hours. The IT period was initially scheduled to run through March 2000. IOT&E, using only 9, would operate out of Kirtland AFB but with testing at other locations. The SIRFC testing was to be performed on the Edwards, China Lake and Nellis AFB ranges.

Initial testing found the CV-22B modifications did not significant degrade baseline V-22 handling qualities, but later work revealed a significant climb performance decrement in some parts of the envelope from added drag. A few MMR development sorties were flown on 7 before all Ospreys were grounded in December 2000. The two CV-22 test aircraft had flown 90 flights for 150.0 hours. Aircraft 7 then had a total of 244 flights and 498.3 hours, and 9 with 152 flights and 338.7 hours.

Testing the EW installation on 9 was a long and complex undertaking. Three test periods in Edwards' huge anechoic chamber allowed optimization of antenna installation and the addition of radar absorbent material in areas around the antennas to reduce signal reflection. The work also included initial SIRFC evaluation and integration of an interference canceller that allowed the many antennas on the aircraft to perform well in close proximity. SIRFC had run into developmental problems and fell behind schedule. The Army decided not to field it for many years, leaving the Navy and USAF to push the system along. The flight test delay imposed by 18's accident allowed the developers to make considerable strides, and most of the system's capabilities were expected to be functional by IOT&E and IOC.

Because of a tight schedule, the CV-22 testing aimed at clearing capabilities for initial deployment. Apart from SIRFC, this also meant some TF capabilities that would remain to be cleared in later testing. Deferred were the lowest TF altitudes plus flight in inclement weather and icing.

The issue of revised AR probe length was finally addressed in 1999 as the CV-22 modification was coming together. The design of the new AR probe went through several iterations. Analyzed were extensible units and probes with a 'dogleg' kink. Developers finally settled on a fixed, straight unit of about 9-ft (2.7-m)

Top: **The need to eliminate V-22 shipboard operating restrictions saw funded research at many facilities in the country. Here a working scale model of the Osprey is moved about a matching scale model of the ship within a low-speed wind tunnel.** NASA

Left: **Aircraft 7 approaches the Edwards AFB enormous main runway in the California desert during its first flight in 19 months, 11 September 2002. Flight with the cargo door raised will be common in V-22 operations to improve crew chief field-of-view.** AFFTC

length and with a greater diameter for sufficient stiffness. This was first flown on aircraft 10 in late 1999. While solving the visibility problem, the length would be a hindrance to shipboard movement. However, the Marines would only fit the probe as a kit for ferry or special missions, requiring about an hour installation time. Both 7 and 9 were fitted with the new probe, but initial MMR testing found it fell within the radar field-of-regard during starboard turns, risking erroneous fly-up commands. However, planning for a fully retractable, telescoping probe was already underway. It would extend to 10.99ft (3.35m) but would retract to a mere stub. The program committed to this design in 2000 and a dummy of the exposed portion was installed on the CV test birds.

Right: **The USAF flight engineer, outside on intercom, and a ground crewman monitor engine start-up preparatory to a test or training flight. Aircraft 7 was given a unique Osprey nose logo.**

Below: **A vital element of the CV-22 testing was optimizing antenna installations for the best EW system performance. This was achieved through many weeks of trials with the aircraft 9 suspended in Edwards' huge anechoic chamber.** Both AFFTC

Because of delays and fleet groundings, the CV-22 program was well behind schedule when the much-inspected aircraft 7 resumed flying on 11 September 2002. TF flight was first demonstrated in April 2003 and early results were very good. It had been decided that 9 would be further modified with some aspects of future CV-22B upgrades. This saw the aircraft grounded from 18 September 2000 to 14 July 2003. It was to undergo further modifications in the summer of 2004 to a final production configuration. Testing increased slightly as some work was shifted to Edwards from the heavily taxed MV-22s. It appeared the project would run through Spring 2006 when operational testing would begin using ship 9 and two LRIP Production Representative Test Vehicles (PRTV). To relieve the schedule and training pressures, and the risk of so much work only with 9, MV-22B aircraft 25 was selected as an Additional Test Article for CV-22B modification in Philly beginning in September 2003 to join the Edwards team in November 2004. Although initial plans were to restore the aircraft to MV-22B standards, but it could possibly be adopted by the USAF in its CV configuration.

The need to conduct additional testing to give AFSOC full performance capabilities, and to evaluate upgrades already planned, made it increasingly likely 7 and 9 would remain Edwards CV testbeds until perhaps 2010. Aircraft 7 might be retired after 2006 and replaced with a production CV-22B. It was expected 8 and 10 would be retired in about the same timeframe.

BHT-GTA
V-22
FIXTURE NO. 1

Production and Service

Photographs on the preceding page:

Top: **The interior of the V-22 wing upper surface shows the integral skin and stringers in the one-piece composite structure.** Jay Miller Collection

Middle: **The first V-22 wing, for the GTA, is being fitted into a manufacturing fixture. Visible is the right-hand tip closeout rib for the torque box and the fixed trailing edge panels over the cove ahead of the flaperons (not installed).** Jay Miller Collection

Bottom left: **The FSD Static Test Article fuselage nears completion in 1988 at Boeing's Ridley Park, Pennsylvania, assembly site. Note the cabin ditching hatch opening atop the aft fuselage.** Tony Landis collection

Bottom right: **Enormous jigs were integral to the V-22 FSD manufacturing process. In the foreground, ship 4 fuselage assemblies have been joined in the late summer of 1988. Beyond is the lower lobe belly skin of ship 5.** Author's collection

The $1.454 billion contract for the first three LRIP lots was signed in June 1996. The release of $42 million for Lot 1 long-lead procurement was approved based on the limited operational assessment during risk reduction testing of the modified FSD aircraft with numerous operational limitations, and very limited flying on EMD's aircraft 7. Criteria for the $402 million Lot 1 (5 MV-22Bs) approval in April 1997 were ferry of 7 to Pax, meeting the empty weight, and flying to 220kts. The 1998 Lot 2 approval (5 MV-22Bs) was predicated on delivery of the next two EMD aircraft to Pax and completion of a specific portion of the static loads tests. Some opined that these represented largely meaningless milestones and inadequate data from overly artificial tests. However, it was necessary to built production-representative airframes for OPEVAL and initial training. Construction of the next 19 MV-22s (9 in Lot 3 and 10 Lot 4) was approved in January 2000.

Construction of aircraft 11 and 12 assemblies began in May 1997. Final assembly of both began at Plant 6 in late 1998. Ship 11 flew on 30 April 1999, was official delivered to the USMC on 14 May, and ferried to Pax on the 27th. Assembly then shifted to Bell's $40 million Tiltrotor Assembly Center at Amarillo International Airport, Texas. Opening in Spring 1999, it enclosed 200,000ft^2 (18,580m^2). Aircraft 13 was the first assembled in Amarillo, and delivered in December 1999.

Boeing performs about 50% of production work, building the fuselage in Ridley Park. This is done from the floor beams up as a forward fuselage (avionics cabinets, fully appointed cockpit and nose), mid fuselage (cabin and sponsons), and aft fuselage (ramp sill to tail). These three sections were spliced together. The fuselage is flown to Amarillo, typically by C-17 or C-5. The wings, proprotors, and associated components come in from Bell's Dallas/Fort Worth plants. It takes roughly two years to build an Osprey. The engines continue to be furnished by the government. Allison became part of Rolls-Royce in the first half of

Below: **The assembly hall in Ridley Park, in August 1988. The fuselage of ship 2 is about 90% complete and 3 (foreground) is still awaiting its empennage.** Author's collection

1995. In 2000 the T406-AD-400 was renamed the AE-1107C Liberty.

In October 2000 the Amarillo plant was very busy with the late LRIP. Aircraft 19 was approaching its delivery date and Bell was beginning its plans to double the assembly building area in 2003, and a flight hangar was being considered. The work force there was expected to increase from the 248 employees at the end of 2000 to 1,500 people by 2008. Likewise, Boeing's Philadelphia site more than doubled from 135,000ft^2 (12,540m^2) in a renovation and relocation completed in June 2003.

The first years of flight test and training saw many short fleet groundings to address uncovered deficiencies or to introduce minor improvements. While common with new aircraft, this always attracted public attention and more statements in the press that the aircraft was not measuring up to expectations and its high cost. One of the more serious such events occurred in late August 2000 when, following a precautionary landing, inspection revealed a loose coupling in the interconnect drive shaft. This was quickly remedied, but it highlighted the many changes expected over the first few years of production. For example, some 86 changes were introduced through modifications within a year of turning MV-22s over to OPEVAL.

EMD ran far beyond the 1998 target. But, as this phase was coming to an end in 2000, the future of the V-22 appeared bright. Recent military operations demonstrated that the aircraft would greatly enhance the America's ability to respond to unconventional warfare or peace-keeping missions where rapid positioning and resupply of highly mobile forces was paramount. Congress had appropriated $1.3 billion in January 2000 for FY01 to procure MV-22s at full-rate, plus long-lead procurements for CV-22s.

The 360 MV-22s, 50 CV-22s, and 48 HV-22s represented $37.3 billion in production, support, and upgrades. The program formulated plans to acquire 16 MVs and 4 CVs in FY01's Lot 5, plus 18 MVs and several CVs in FY02's Lot 6, with first delivery in 2003. The CVs were to peak at 2-5 per year. The USAF paid for the common airframe while SOCOM funded CV-unique content. The Marines were considering converting part of their order to CVs. They sought to buy a maximum 30 MV-22Bs per year, although desiring to raise this to 36 in order to field the aircraft more quickly and to reduce overall costs. Navy HV-22 production was to begin in 2010, possibly raising the overall monthly rate to as many as 42. All these quantities and out-year planning numbers fluctuated as budgets were adjusted.

Unit costs continued to be a major issue. At the beginning of EMD estimates varied between $40 and $57.5 million, with uncertainties at that stage to be expected. The Navy's target was $45-47 million, but this appeared increasingly unrealistic by Spring 2000. The CV-22 cost rise was the greatest. This prompted a reduction in planned annual buys and peak production rate down to 24, stretching-out the buy. Unfortunately, this would see the badly aged CH-46 serving on through 2015. The CV purchase would extend to 2013. Although keeping annual costs in check, it did not help reduce unit costs to original estimates. By the end of 2000 a dispute arose during negotiation when the manufacturers presented a much higher cost, reported to be about $66-70 million.

The decision to proceed to FRP was targeted for 27 November 2000. The first FRP contract was expected to be signed in March 2001, worth $1 billion.

The first unit to received production MV-22s was VMMT-204, created at MCAS New River on 10 June 1999. They would take on the first dozen machines and 20 instructor pilots as the sole V-22 aircrew and maintainer training unit for both USMC and AFSOC. Their first aircraft were the four machines completing OPEVAL. IOC was set for January 2001 with 12 aircraft.

Top: **The Full Flight Simulator is built by FlightSafety International and possesses a motion base and high quality out-of-window vision. It is an essential part of Osprey pilot training.** Bell Helicopter

Left: **The V-22 is the bus that takes the grunts to war. As evident here, comfort is not a priority. The crew chief will be bumping knees moving about the cabin.** DoD

Right: **The cargo hold of the V-22 prototype during pre-flight maintenance at Bell's Arlington Municipal Airport facility.** Jay Miller

Below: **The view from the right pilot seat as the V-22 AR probe is latched into the hose basket coupler of a Marine KC-130 tanker.** Ron Culp

At the end of 2000 the USMC adopted a curriculum for student pilots destined for the MV-22. The first class of four Marines was to undergo the training in FY01. Students would initially follow the established helicopter primary training in the fixed-wing Beech T-34C Turbo Mentor and the Bell TH-57 (JetRanger) helicopter. Their Joint Advanced Tiltrotor Training would then consist of 340 flight hours in the Beech TC-12B (King Air) airplane at NAS Corpus Christi, Texas, and further rotorwing work in the TH-57B/C at NAS Whiting Field, Florida. The classroom, simulator and in-flight studies would focus on multi-configuration instrument flight, low-speed handling, shipboard landing qualification, confined area landing, high-low penetration tactics, and instrument-to-visual navigation. The students would then receive their wings before moving on to VMMT-204 where 120 hours of academics, 65 hours sim time, and 40 flight hours were planned to qualify an MV-22 pilot.

The USMC announced plans in early 1999 to place the Ospreys in 21 regular and 4 reserve units, 12 aircraft each, by 2015. They would replace CH-46s and early CH-53Ds, and displace CH-53Es to other units. The initial four combat outfits would be East Coast squadrons of the Second Marine Air Wing (MAW), New River. The first was to be Medium Helicopter Squadron HMM-264, the Black Knights, beginning their transition in January 2001. Following training, it was to join a Marine Expeditionary Unit in the fall of 2001 and deploy for the first time six months later. Their full complement of aircraft was to be reached in 2005, although the first sea deployment was planned for 2003 aboard the USS *Iwo Jima*. Four West Coast squadrons of the Third MAW and Marine Air Group 16, based at Miramar, California, would follow. They would be fully equipped with Ospreys between 2005 and 2007. Three squadrons of the First MAW in Hawaii, Kaneohe Bay, would replace their CH-53Ds with Ospreys between 2007 and 2012. Two 46 units stationed at MCAS Futemma, Japan, were to transition beginning in 2012. After this three more units at New River and then three Sea Knight squadrons at Camp Pendleton, California, would convert through 2009. Two more CH-56 units of the First MAW were to take on MV-22s in 2012. Four reserve outfits were also to convert in 2011 and 2012. The first would be a 46 unit at NAS Norfolk, Virginia,

Test Aircraft Summary

Aircraft No.	Bu No.	C/n	First Flight	Disposition	Notes
1	163911	90001	19 March 1989	museum display	FSD flight test
2	163912	90002	9 August 1989	live-fire subject	FSD and Risk Reduction flight test
3	163913	90003	6 May 1990	museum display	FSD and Risk Reduction flight test
4	163914	90004	21 December 1989	crashed 20 July 1992	FSD flight test
5	163915	90005	11 June 1991	crashed on first flight	FSD flight test
6	163916	90006	never flown	parted-out and used for live-fire testing	scheduled FSD flight test
7	164939	90007	5 February 1997	modified for CV-22B testing	EMD flight test
8	164940	90008	23 August 1997	follow-on testing at Pax River	EMD flight test
9	164941	90009	17 July 1997	modified to first CV-22B	EMD flight test
10	164942	90010	16 January 1998	follow-on testing at Pax River	EMD, was to be retired after EMD

followed by two CH-53E squadrons at Edwards AFB, and the last at NAS Willow Grove, Pennsylvania. By mid-2000 the USMC was seeking to place nine of its proposed California MV-22 squadrons in a single location in the state, Edwards being one site considered.

The first four CV-22s, including aircraft 9, were to go to the 58th Training Squadron at Kirtland AFB, New Mexico, in 2003. IOC was set for September 2004 with six aircraft assigned to the 16th Special Operations Wing, Hurlburt. A squadron was expected to be established in the Pacific region, at Osan Air Base, Korea, or Kadena AB, Japan. This would be followed by a unit at RAF Mildenhall serving Europe.

Above: **Quick maintenance aboard USS *Saipan* gives a good look at the proprotor pendulum absorber masses protruding from the spinner.** DoD

Left top: **Caught in the moment of conversion and acceleration as the landing gear finishes its retraction cycle, the Osprey is noteworthy in having a level deck angle where a helicopter would have a severe nose-down attitude.**

Left: **A heading change as the Osprey goes feet-dry crossing the beach.** Both Ron Culp

All the production and fielding plans were suspended following the crash of 18 and the long period of redesign and evaluation that followed. When deployment planning was again solidified, the same general scheme was likely to be adopted. The grounding left VMMT-204 with eight MV-22s, 11-13, 15-17, and 19-20. The new 21-22 were then ready for flight at Amarillo, seven were in final assembly, and 15 aircraft under fabrication.

The plans for production beyond 2000 were immediately revised pending any future full-rate decision. The Blue Ribbon Panel had recommended that LRIP continue at the lowest practical rate. Keeping the production line 'warm' was vital in retaining knowledgeable, skilled labor, the commitment of subcontractors and vendors, and keeping overall production costs down. It was understood that the aircraft being manufactured would require modifications later, at some expense.

Contracts valued at $1.5 billion for 9 MV-22s in FY01, and 9 MV-22 (delivery through 2004) plus 2 CV-22 PRTVs (delivery through 2005) in FY02, were let as modifications to existing contracts. The pressures on the CV-22 testing and training schedule had prompted earlier initiation of CV construction, with IOC expected in October 2009. As 11 machines was considered the lowest practical annual number, another LRIP contract for 11 Lot 7 machines was let in early 2002 at $770 million. Beyond this, $1.8 billion was programmed for 9 MVs and 2 CVs in FY04, including hundreds of millions more for continued R&D, and $1.5 billion for the same in FY05. Billions were being appropriated for the program with an uncertain future. Until late in 2002, aircraft continued rolling off the Amarillo line that could not be flown or delivered. This began to tax the ability to store them within the available space, expansion of the facility having been suspended.

With encouraging development and testing progress, the V-22 got a boost in May 2003 when the skeptical Pete Aldridge, Under Secretary of Defense for Acquisition, Technology and Logistics, declared his satisfaction with the program's progress. By summer 2003 the program was urged to increase LRIP from 11 to 15 aircraft annually beginning in FY05 vice FY07, and climbing to 41 by 2009, as a means of reducing unit cost through economy of scale and hastening deployment. However, the desire to ensure funds remained for development and upgrades kept the FY05 number at 11, with plans for 17 in FY06, and increases of 50% each year thereafter. This could be critical in meeting the general objective to slash unit cost by $10.7 million to about $58 million by 2010. A CV-22 was expected to cost about $65.8 million.

A full-rate production decision was expected in October 2005 following the second round of OPEVAL. With more than 50 aircraft produced by the time, program cancellation would seem doubtful. Indeed, Marine Corps IOC was planned for September 2004 and fleet deployment in December 2005.

Production Aircraft Summary

Year Funded	Type	Bu No / c/n	Delivered	Disposition	Notes
1997 LRIP Lot 1	MV-22B 11	165433/90011	25 May 1999	VMMT-204	initially OPEVAL test asset
	MV-22B 12	165434/90012	1 Nov 1999	VMMT-204	initially OPEVAL test asset
	MV-22B 13	165435/90013	13 Nov 1999	VMMT-204	initially OPEVAL test asset
	MV-22B 14	165436/90014	17 Jan 2000	crashed 8 April 2000	initially OPEVAL test asset
	MV-22B 15	165437/90015	12 Mar 2000	VMMT-204	initially OPEVAL test asset
1998 LRIP Lot 2	MV-22B 16	165438/90016	30 July 2000	VMMT-204	
	MV-22B 17	165439/90017	7 Jul 2000	VMMT-204	
	MV-22B 18	165440/90018	21 Aug 2000	crashed 11 Dec 2000	initially VMMT-204 training asset
	MV-22B 19	165441/90019	15 Oct 2000	VMMT-204	
	MV-22B 20	165442/90020	6 Nov 2000	VMMT-204	
	MV-22B 21	165443/90021	11 Oct 2002	VMX-22	initially Pax test asset
	MV-22B 22	165444/90022	16 Jan 2003	VMX-22	initially Pax test asset
1999 LRIP Lot 3	MV-22B 23	165837/90023	9 Jul 2003	VMX-22	initially Pax test asset
	MV-22B 24	165838/90024	1 Apr 2003	VMX-22	initially Pax test asset
	MV-22B 25	165839/90025		AFFTC	to be remanufactured as CV-22B test asset
	MV-22B 26	165840/90026			
	MV-22B 27	165841/90027			
	MV-22B 28	165842/90028			
	MV-22B 29	165843/90029			
2000 LRIP Lot 4	MV-22B 30	165844/90024			
	MV-22B 31	165845			
	MV-22B 32	165846			
	MV-22B 33	165847			
	MV-22B 34	165848	17 Jul 2003		first with production Block A changes
	MV-22B 35	165849			
	MV-22B 36	165850			
	MV-22B 37	165851			
	MV-22B 38	165852			
	MV-22B 39	165853			
	MV-22B 40	165956			late addition, out of sequence BuNo
2001 LRIP Lot 5	MV-22B 41	165940	late 2003	VMX-22	first to VMX-22
	MV-22B 42	165941			
	MV-22B 43	165942			
	MV-22B 44	165943			
	MV-22B 45	165944			
	MV-22B 46	165945			
	MV-22B 47	165946			
	MV-22B 48	165947			
	MV-22B 49	165948			
2002 LRIP Lot 6	MV-22B 50	165949			
	MV-22B 51	166383 (165950 ?)			
	MV-22B 52	166384 (165951 ?)			
	MV-22B 53	166385 (165952 ?)			
	MV-22B 54	166386 (165953 ?)			
	MV-22B 55	166387 (165954 ?)			
	MV-22B 56	166388 (165955 ?)			
	MV-22B 57	166389 (165956 ?)			
	CV-22B W5	020024/91005 †			
	MV-22B 58	166390			
	CV-22B W6	020025			
2003 LRIP Lot 7	MV-22B 59	166391	31 Dec 2004 *		
	MV-22B 60	166392	31 Jan 2004 *		
	MV-22B 61	166393	28 Feb 2005 *		
	MV-22B 62	166394	31 Mar 2005 *		
	MV-22B 63	166395	30 Apr 2005 *		
	MV-22B 64	166396	31 May 2005 *		
	MV-22B 65	166397	30 Jun 2005 *		
	MV-22B 66	166398	31 Jul 2005 *		
	MV-22B 67	166399	31 Aug 2005 *		
	MV-22B 68		30 Sept 2005 *		
	MV-22B 69		31 Oct 2005 *		
2004 LRIP Lot 8	MV-22B 70				
	MV-22B 71				
	CV-22B W7	040026			
	MV-22B 72				
	MV-22B 73				
	MV-22B 74				
	MV-22B 75				
	CV-22B W8	040027			
	MV-22B 76				
	MV-22B 77				
	MV-22B 78				

The rest are undefined

* planned; † CV-22B assigned USAF tail numbers are xx0024 through xx0073, where the xx will be the last two digits of the fiscal year contracted. Corresponding production numbers were supposed to run from 91001 through 91050.

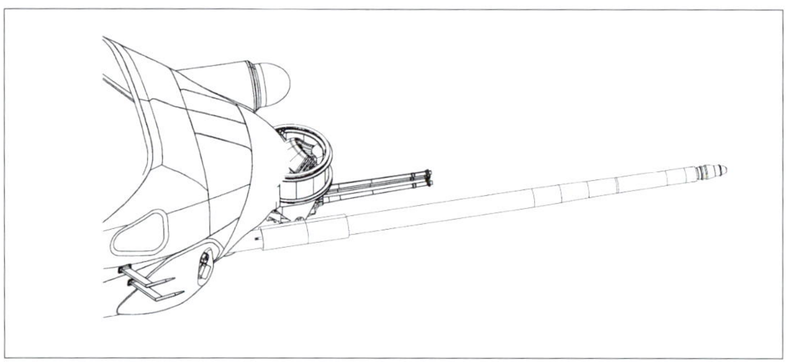

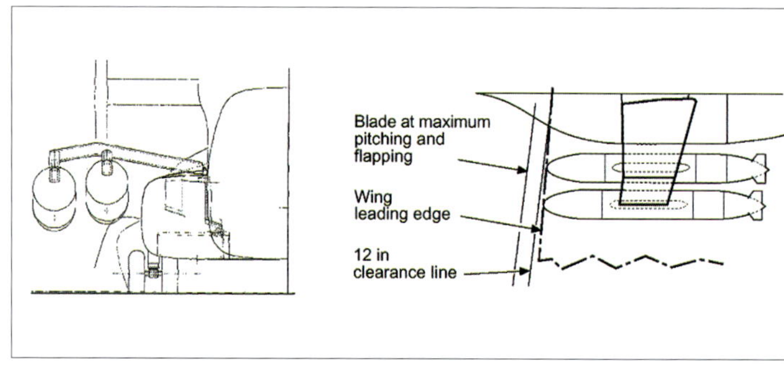
Blade at maximum pitching and flapping
Wing leading edge
12 in clearance line

UNITED STATES OF AMERICA

Chapter Six

The Future

Follow-On Development

Many potential improvements and additional capabilities were identified during the course of V-22 development that could not be immediately incorporated because of budget or schedule constraints. Fortunately, research and development funding continued to be authorized by Congress after the end of EMD. However, these were limited resources compared with the many potential fixes and improvements. The program's cost cap meant slow progress on the design and testing of upgrades. By late 2000 some 149 changes had been identified and 122 were being actively pursued. The aircraft would see upgrades, modifications, and testing for years to come. In some cases the continuing development would be reversing compromises made at the beginning of EMD to reduce aircraft weight and cost but which proved operationally ill advised.

Photographs on the preceding page:

Top left and right: **One concept for the nose gun installation had the turret at the end of an extended nose fairing. This artist's concept also shows the extendible AR probe.** Author's collection

Upper left: **This drawing shows the fully retractable AR probe at its extended length and a notional gun installation. Mounting the gun in an extended nose turret would improve field-of-fire and eased integration issues. However, proximity to the CV-22 radar seen here raises vibration concerns.** Author's collection

Upper right: **External tanks and ordnance has been an open question since the earliest day of the JVX. This concept drawing shows how proprotor arc is addressed, but jettison dynamics in the complex and varying tiltrotor flow field would be a tougher nut to crack.** Bell Helicopter

Lower left: **A Bell tiltrotor gunship proposal from 1982.** Bell Helicopter via Ned Gilliand

Lower right: **A Bell commercial tiltrotor proposal from 1991.** Bell Helicopter via Ned Gilliand

Bottom left: **A Bell/DARPA/Army/Marine advanced, low-observables configured gunship proposal from 1988.** Bell Helicopter via Ned Gilliand

Bottom right: **A Bell VV-22 proposal optimized for the Presidential fleet from 1988.** Bell Helicopter via Ned Gilliand

This page, right: **One alternate mission for the Osprey that has been pursued with some vigor is aerial refueling. The speed range of the aircraft would make it compatible with a wide variety of receivers.** Author's collection

Planning for Follow-on Test and Evaluation began in earnest in 2000. Much work remained from EMD that simply continued, with some capabilities remaining to be implemented and evaluated. These included identified avionics upgrades, a full evaluation of sling load configurations throughout the speed range, the long-planned cabin auxiliary fuel tanks, complete clearance of AR, and work on the cabin over-pressurization capability. Efforts continued on removing remaining flight restrictions and even further opening the operational envelope. There were many undesirable or unimplemented aspects of the cargo handling and aerial delivery system to be addressed over the coming years. NAVAIR built a functional mockup of the cabin at Pax to facilitate that work. The balky rescue hoist was to be addressed with tests of an electrical unit in July 2004. Of greater concern was that the forward door was too narrow to safely hoist aboard a survivor seated on a foliage penetrator or in a Stokes litter at the end of the cable. This problem would not be easily remedied.

Responding to the ship 14 and 18 crashes took precedence, but also offered an opportunity to incorporate fixes judged prudent or complementary. Congress made known its concerns that the V-22 did not have a suitable flight data recorder and cockpit voice recorder, and that some manner of warning of impending VRS should be incorporated. Resolving safety, reliability, maintainability, and affordability issues impacting operational suitability, cost, and weight remained paramount until a FRP decision. Block Bs and C changes were being programmed as flight testing resumed, but many operational enhancements in pre-planned product improvement (P^3I) would only crystallize in future years. The volume of work would see test teams maintained at Pax River, New River, and Edwards for many years.

Among the systems being addressed for potential revision was the fuel system that, lacking positive pilot control of pumps and valves, had proven operationally cumbersome. Fuel cells that failed to meet crashworthiness standards were to be retrofitted with those from a different manufacturer. The fuel dump line below the aft portion of the aircraft was in a flow field that could carry fuel onto the aircraft and potentially into avionics cooling inlets. The line was to be moved or replaced with one that dropped down. To aid in cg management, the starboard aft sponson fuel tank would be deleted and fuel cells added in the wings, actually increasing overall capacity. Many operators felt the V-22 would greatly benefit from addition of a true head-up display (HUD). Slow trimming and the ability to stall the trim motor meant a residual pitching moment remained during conversion and reconversion. Climate control also required improvement as cabin temperatures could not be easily maintained at comfortable levels. The NBC kit for the environmental control unit was abandoned when the requirement was deleted. The kit had suffered developmental problems and it was proving difficult to achieve an adequate pressure differential to prevent hazardous agents from breaching the interior. In any event, during combat the ramp would be dropped and the cabin contaminated. Instead, the crew and passengers would rely on personal protective gear.

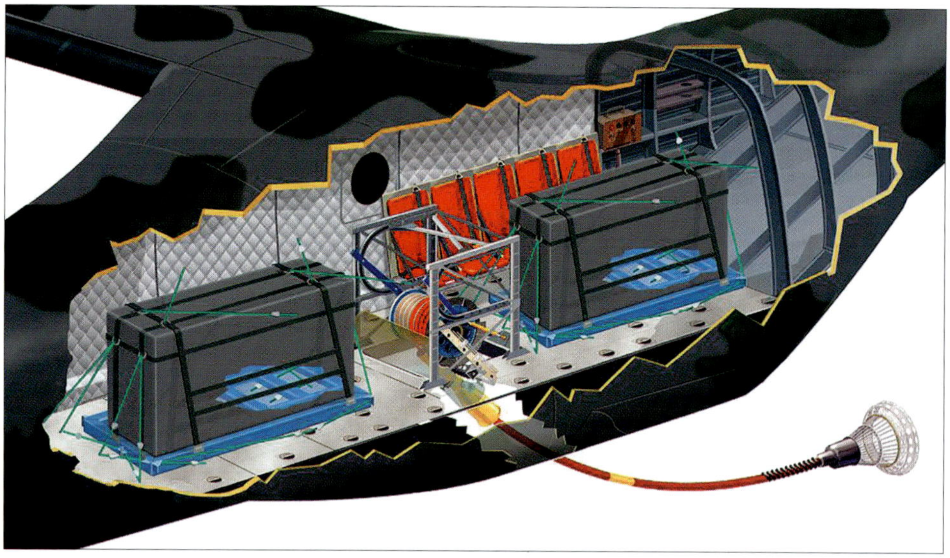

Left: **The tanker mission might be accommodated with minor modifications and roll-on equipment. This concept shows two cabin aux tank and a reel assembly feeding the fuel hose out the aft cargo hook 'hell hole'.** Bell Helicopter

The cost cutting that accompanied the CV-22 proposal saw elimination of some equipment AFSOC had sought, including a low probability of intercept (LPI) radar altimeter (RADALT) and a survivor locator system. A new RADALT, possibly incorporating LPI, was addressed after testing revealed false readings from signal reflection off sling loads. The altitude limit of the RADALT was also undesirable for the CV mission. The lack of an anti-skid was very inconvenient during heavy weight run-on landings, which the USAF expected to do frequently. The crew had to apply brakes gingerly, but flat-spotted and blown tires were experienced. A backup braking system that used a battery-operated pump was also being testing on ship 9 and others.

During OPEVAL a .50 caliber machine gun was installed at the end of the ramp on one aircraft in an improvised and non-operational mount. Although limited in field of fire to the sides, as offered by waist guns, the ramp gun at least did not risk hitting the nacelles or proprotors and so was an easy installation. Clearly the operators sorely felt the lack of a gun. Combat experience demonstrated that a gun was essential in suppressing enemy ground fire during an assault or rescue. Early concepts had gun barrels emerging from the nose tip fairing, but this eventually evolved into a chin turret. Concern arose over vibration problems for the CV-22 radar and restricted fields of fire to avoid hitting the AR probe, proprotors, and nacelles. Plans included slaving the gun to the FLIR and helmet-mounted NVG/HUD in addition to operating the gun using the track handle in the center console. At one point during 1999 the GAU-19/A three-barrel .50 cal weapon was selected and integration design begun. The weapon was to be chin-mounted ahead of the FLIR turret, firing at 1,200 rounds per minute and weighing up to 300 lb (136kg, sans ammunition). It was to have a 750-round linkless feed system under the floor that could be reloaded in flight. In 2000 the development was suspended for a few years because more urgent matters consumed projected funding. The ramp gun was a likely interim solution.

Liquid crystal displays (LCD) were being used for several small V-22 cockpit displays during EMD. It was well into flight test that LCDs became practical for the large MFDs. The cost and weight savings of replacing the large MFDs with LCD flat panels made the substitution desirable. The new units took up considerably less volume, were easier to read in daylight, reduced annoying MFD glow during night operations with NVGs, and were more reliable. The flat panels were introduced into the production line in Lot 2 with aircraft 19 while earlier aircraft were to be retrofitted. Another such weight-savings improvement was a lightweight exterior paint with superior IR signature reduction properties. It was first applied to aircraft 22 for testing.

P³I focused on MV and CV capabilities or subsystem changes to meet emergent threats or to realize a short-term and cost-effective improvement with little additional investment. Both customers saw the need to increased engine power to provide a reserve under some conditions where the baseline aircraft exhibited marginal performance. Both also needed to upgrade their communication and navigation systems to introduce a Traffic Collision Avoidance System compliant with the Global Navigation System/Global Air Traffic Management improvements being mandated worldwide. Advanced radios and data- links would be major portions of this effort, in addition to new tactical datalinks possibly in Block B. A Ground Collision Avoidance System was also on a long 'wish list'.

The complex CV-22 mission meant SOCOM was always seeking to incorporate the latest technology. Among the most notable desires were radar upgrades to include auto-TF, MMR radome deicing, and coupled hover capability. In 2001 attention turned to installing the AAQ-24(V) Directed Infrared Counter Measures system that was then reaching maturity. This directed a laser at approaching IR-guided missiles to cripple their seekers. The change would see as a pair of turrets under the aft portion of the sponsons. Aircraft 9 was given dummy turrets for an early look at the performance decrement during CV-22 flight testing. Some or all of these changes would be included in a Block 10

Bottom: **Phased array radar elements within the top fairing eliminate proprotor interference problems in this AEW concept. However, mounting the assembly to the wing center-section while also allowing ready access to the components in this area could prove problematic.** Bell Helicopter

Right: **The first tiltrotor UAV, the Bell Boeing Pointer, rises to a hover on its maiden flight. The poles mounted beneath the aircraft are to reduce motion rates and prevent excessive angles close to the ground. Bob McClure collection**

configuration. Other Block 10 changes included addition of a radio control head and MFD with custom display in the center console, within easy reach of the flight engineer in the jump seat. Aircraft 9 and the first production lot of four CV-22s became Block 0. Elimination of the MV-22's laser-warning receiver was a significant operational impediment hoped to be rectified with a future system, perhaps in Block 20. It was expected that Block 20, to include the nose gun, would be introduced before the full production run of CVs. Earlier aircraft would be eventually brought up to the latest standard. Further hardware changes and the inevitable software revisions were expected to generate further upgrades about every three years. Farther out, the USAF hoped to eliminate the MMR and many other emitters, substituting ultra-precision navigation and LPI comm systems, to improve covertness.

With the TF radar, AR probe, and EW antennas, the CV-22B continued the reputation of ugly SPECOPS aircraft. But the weight and drag impact were a matter of concern. Even with all the 'warts', developers were optimistic that the aircraft would meet its 500nm range and exceed its cruise speed requirement. However, the weight and drag addition from Block 10 gear and the gun in Block 20 would almost certainly make this impossible. Although the manufactures continued looking for weight reduction, the USAF considered altering its requirements and ops concept, perhaps even reducing the assault team to 12 from 18.

V-22 VSTOL performance was less than originally predicted, and operation in such recent conflicts as Afghanistan would have been significantly impacted. These concerns brought attention to more fundamental improvements in tiltrotor technology. In an effort to further reduce pilot workload, consideration was being given to introducing the XV-15's discrete nacelle settings. Normal practice in reconverting to helicopter mode was to go to 60° nacelle angle, then 75°, and then 95° with appropriate power settings for each. This process requires care in setting a rotation rate with the switch and then cross-checking nacelle angle indications to stop at the desired angle. Discrete settings would simplify transition and allow attention to be diverted to other tasks. Automated STO mechanization, possibly with scheduled nacelle and pitch commands as a function of ground speed, would greatly improve short-field performance and benefit CV-22 operations.

In 2001 Boeing proposed introducing shape memory alloy actuators in the proprotor blades. This would see a material built into the blade that changed shape when an electrical change was applied. It would twist the blade as a function of the charge to a configuration more suitable for individual flight conditions throughout the mission profile. This optimization held promise of significant performance improvement, including a 10% increase in payload and 15% extension to mission radius, all achieved at much less cost than uprating the transmission. Reduced noise and vibration levels were also possible. An alternative was a blade leading edge slot being explored by Bell.

Another idea under research in 2000 was active flow control over the flaperons to prevent separation at high deflection angles in hover and conversion. A device would eject pulsating jets of air at the point of flow separation across the flaperon span. Flaperon settings up to 85° (72.5° the normal maximum) would provide additional download relief that could add as much as 1,000 lb (450kg) of payload or more hover margin. The blown flap approach was tested on the XV-15 during the summer of 2003 with 14% download alleviation realized. Another device to reduce download was tested on the XV-15 in 2000. This was a 'butterfly' or twin, angled panels projecting a few feet above the wing at the fuselage center to reduce the lift lost through the fountain of air generated there in hover that recirculates through the rotors.

Other Operators, Other Missions

In the heady days after the JVX contract was awarded it seemed that the V-22 would be bought to fill many roles in numbers exceeding 1,000. But, the planned buy was much reduced during the extended development and price increases. Many advocates expected purchases beyond the immediately planned numbers, but the Osprey needed to prove itself through years of service before winning over skeptics who held the DoD purse strings.

Even before FRP, some hopes for expanded sales were dashed. The long-contemplated Navy HV-22 remained notional. No money was ever programmed in budget projections for the aircraft to help reduce unit cost. It was further undermined in early 2001 with adoption of the SH-60 Seahawk to replace HH-3s and HH-46s. The service stated it probably would not be buying the Osprey for the foreseeable future because it did not fit mission requirements. The

Bottom: **The Eagle Eye is to see service with the US Coast Guard as a shipboard UAV. It is shown here in early Navy testing at the end of its gear cycle as it converts to forward flight.** Bell Helicopter

Left: **Bell Boeing conceived a number of commercial derivatives of the Osprey in the early years of JVX development. These began with minimally revised aircraft like those shown here with a cargo carrier firm.**
Jay Miller Collection

Bottom: **One of the EUROTILT concepts for a 12-19-passenger tiltrotor commercial aircraft tilted just the rotor hub and proprotors.**
Author's collection

potential for the SV-22 reemerging also dimmed. With the end of the Cold War and dramatic reduction in the Russian submarine threat, the S-3 was to be retired and surface ships re-equipped with SH-60s. In 1999 the USAF was contemplating replacing its 105 MH-60G Pave Hawk helicopters in the CSAR role by 2007 with CV-22s. However, by late 2001 budgetary pressures had made such an acquisition impractical. The Pave Hawks were to be upgraded and retained for many more years. Any formal competition for a replacement would not begin before 2006, with fielding by 2015. The USAF appeared to favor the S-92 or EH101. Although many states' Guard units expressed a desire for the versatile and high-performance V-22, it would likely prove too costly.

One of the originally identified potential missions of the MV-22 was to replace the VH-3D Sea King as the 'Marine One' vertical lift transport for the President of the United States. In 2003 the Navy launched its effort to field a replacement by 2007-2008 under the VXX program, with the first exploratory contract expected in early FY04. The speed and range of a 'VV-22' would be most welcome in the mission. But, the delayed development and production of the type put it out of the running, with the S-92 and EH101 chosen as leading contenders.

The Army remained a source of hope for the Osprey manufacturers. However, the service's decision to remanufacture their Chinooks, and other aviation priorities, made any such purchase unlikely for the foreseeable future. The Army's move after 2000 to a lighter, more quickly deployed and mobile combat force had some looking again at the Osprey. The long-term Future Transport Rotorcraft (FTR, formerly the Joint Transport Rotorcraft) program was looking at the V-22 as well as other options to replace the CH-47 beginning around 2015, but this requirement soon grew to a much larger aircraft. The V-22 was also considered a potential platform for an airborne tactical laser weapon to engage ground targets and projectiles in-flight. Limited funds still made any Army purchase appear doubtful.

Even before first flight of the Osprey, Bell Boeing was examining export potential. The team had reached agreement, or was in discussion with over-seas firms to explore sales possibilities in their respective countries. All this was set aside after FSD cancellation. The V-22 was offered in the mid-1980s to fill an Australian requirement, although not selected.

In late 1999 a United Kingdom study found the V-22 possessing a marked operational superiority over helicopters in fulfilling the Support Amphibious and Battlefield Rotorcraft and the Future Amphibious Support Helicopter programs, the latter replacement for fleet airborne early warning helicopters. The UK had a test pilot in early EMD flying and the attitude was one of wait-and-see. But, production delays were placing the Osprey out of the running. By 2000, sales in the European market seemed remote as the new indigenous helicopters, the NH-90 and EH101, filled more and more requirements. Other nations, including Japan and Israel, continued to express an interest. Foreign orders were likely only after the type was well established in US service and actual operational costs become clear.

Upgrades and modifications to the MV-22 and CV-22 were a virtual certainty to improve capabilities and make them suitable for additional missions. The flexibility in filling multiple roles identified by the DoD was one of the original motivations for the program. The manufacturers suggested how this could be done with minimal airframe alterations, especially employing rolling-on and -off kits of specialized equipment. This would easily allow the machine to be used for COD, VOD and executive transport duties.

A Navy and Marines needed an airborne drogue-refueling tanker capable of off-loading as much as 16,000 lb (7,260kg) of fuel. A 'KV-22' with two cabin aux tanks, flying up to 275kts during a 4-hr flight, was contemplated to replace or augment Marine Corps KC-130s. With imminent retirement of the USN S-3 buddy tankers, such an option appeared desirable, albeit expensive. A hose reel assembly would be installed on the cargo ramp or in the center of the cargo deck with the hose and drogue trailing out a cargo hook 'hell hole'. Alternatively, the aircraft could be used to ground-refuel rotorcraft and combat vehicles at a

Right: **Bell Boeing's artist concept for the Quad TiltRotor retains blade fold but a wing stow feature appeared impractical. The size of the machine would ban it from all but the flight decks of the largest carriers, although the manufacturers' showed a notional vessel that could accommodate the aircraft below deck.** Bell Helicopter

Bottom: **Bell pursued the tilt-fold concept over many years. The D-272 design was for this USAF medium transport dedicated to search and rescue.** Jay Miller Collection

Forward Area Refueling Point. By 2001 initial development of the tanker system was under contract, but more urgent matters abounded.

Fleet airborne early warning (AEW) was another proposed mission. This variant would naturally raise concerns with a top-mounted radar antenna rotodome and its compatibility with the proprotors. Phased array radar segments mounted to the airframe could eliminate the need for a rotating assembly and reduce the overall size. A common airframe for the AEW, ASW and standoff EW missions has been sought. Although lost during deep budget and fleet cuts, the 'EV-22' idea still had merit.

Other Tiltrotors

After the V-22 program was well underway, the manufacturers sought to introduce the tiltrotor into other areas of aviation. Derivatives or other designs would take on even more missions, including armed attack. The team developed the 650-lb (286-kg) maximum takeoff weight D-340 Pointer uninhabited air vehicle (UAV) as a demonstration of a remotely piloted tiltrotor for reconnaissance and surveillance. First flying in 21 November 1988 (beating the V-22 into the air), this work eventually yielded the 2,000-lb (907-kg) maximum takeoff weight Bell TR911X Eagle Eye UAV. The aircraft flew for the first time on 10 July 1993. It competed for two Navy contracts before being selected in 2003 for a US Coast Guard requirement as the slightly larger HV-911. It was by far the highest speed but also heaviest VSTOL aircraft in these competitions, yet meeting or exceeding almost all requirements.

The potential market for a commercial tiltrotor had been one factor in the Congressional support for the V-22. Bell Boeing sought to leverage off V-22 experience in conceiving commercial derivatives of the Osprey and larger aircraft. A commercial V-22 could accommodate up to 31 passengers or freight, or 39 in a new, pressurized fuselage. However, market surveys suggested that a small executive transport was the best entry niche for the civil tiltrotor. With Boeing wishing to pursue only large-capacity machines, Bell proceeded alone, initiating design in September 1996. The company sought foreign partners and joined with its long-time collaborator Agusta, of Italy. After ups and downs in concert with V-22 fortunes, the BA 609 design was realized. This was an entirely new machine of 16,000-lb (7,258-kg) maximum GW accommodating six to nine passengers. It brought to fruition Bell's decades-long dream of a civil tiltrotor.

The 609 fostered more than 80 orders even before cutting metal, but was greatly slowed when the V-22 entered its troubled period in 2000-2001. Textron struggled with severe cash-flow difficulties while performing extensive reviews and analysis like the Osprey that the corporation felt had to be seen as successful before the 609 would be commercially viable. After a nearly two-year delay and temporarily halving the number of test vehicles, the prototype finally flew on 7 March 2003 and appeared headed for certification and deliveries in 2007. The aircraft featured composite construction, FBW, and FADEC. Nominal range was to be 750nm and 275kts maximum speed. Significantly, the Federal Aviation Administration (FAA) did not require the 609 to demonstrate autorotation to landing. After all, neither the FAA nor the military required airplanes to perform unpowered landings for acceptance.

The 609 reflected refinements of the tiltrotor and responses to lessons learned from the XV-15 and V-22. The TCL was abandoned for a collective power lever. The wing had single-segment trailing edge flaperons, and the T-tail lacked a rudder. A T-tail reduced PU/SS and elevator buffet induced by proprotor downwash. Directional control in airplane mode is commanded with proprotor differential collective pitch. The aircraft also employs differential collective pitch for lateral control. A full-time attitude stabilization system was to make flying the BA 609 very pleasant. The nacelle control switch commanded discrete nacelle settings of 75°, 60°, and 0° at 3°/sec, with 40 seconds to convert. This is especially useful flying instrument approaches in reducing pilot workload. Typical disk loading was comparable to the XV-15 at 15psf (73kg/m^2).

Military versions of the 609 were considered attractive because the price, estimated at about $15 million, was considerably less than the Osprey. A trainer for V-22 pilots and a gunship version was conceived. A notional 'UV-609' would find utility in many missions, although it would require structural 'beef-up' and likely an aft fuselage ramp, plus military avionics. However, the possibility of buying such machines seemed remote in the DoD budget environment. The USCG was also taking notice of a 'HV-609' to enhance its capabilities. However, when the 609 program stalled in 2001-2002 this opportunity faded.

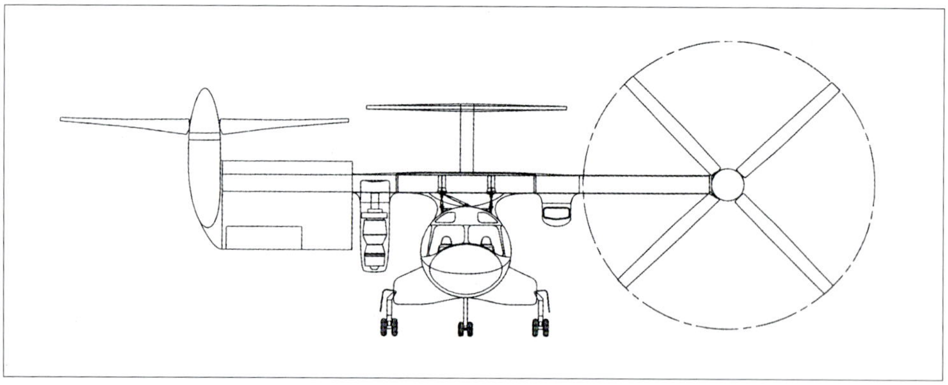

Left: **AgustaWestland's ERICA design envisioned tilting the outboard wing sections under the proprotors to eliminate the majority of the download. The concept also envisioned a comparatively small proprotor diameter that could be operated in airplane mode on the ground.** Agusta via M McCluer

Bottom: **The first Bell/Agusta BA 609, N609TR, during post-flight maintenance at Bell's Arlington Municipal Airport (Plant 6) facility, early 2003.** Jay Miller

Bell pursued the potential of the tiltrotor as a higher-speed aircraft before and after award of the JVX contract. One design proposed using minimally modified V-22 wing/nacelles and possibly empennage on a new fuselage to produce an armed escort/gunship. The pilots/gunners were under a bubble canopy and it had a modest cabin. It featured a chin-mounted gatling gun, top-mounted sensor pod, six Maverick air-to-surface missiles, and two Sidewinder air-to-air missiles mounted externally. This concept had merit in that the V-22's speed made it impractical to fly with attack helicopters for fire-support. But, the 'AV-22' received a cool reception in the late 1980s and did not resurface except as an armed BA 609.

With the tiltrotor apparently proving to be viable and practical, numerous other tiltrotor designs, both military and commercial, were conceived by Bell and Boeing that had little commonality with the Osprey. One that attracted attention was the Quad TiltRotor (QTR) transport with engines and rotors at the tips of two tandem wings. In 2000 the DoD was looking at the feasibility of a heavy-lift VSTOL transport in the class of the C-130. The Army's FTR program looked at tiltrotors as a potential candidate and the Navy COD requirement would benefit from higher capacity than the C-2A. In studies of VSTOL aircraft for these requirements the tiltrotor consistently came out ahead in speed, range, endurance and payload.

In early 1999, Bell Boeing revealed a QTR design that had been under study for about two years as a possible offering in a FTR competition. Bell had first looked at a quad tiltrotor in 1979 as the D-322. The later QTR concept had a cruciform tail, but this was eventually changed to just a vertical tail without rudder. Even this was then eliminated for a tailless design when it became clear the four proprotors provided all the essential directional control. These were all to be connected through a power transmission system for safe control in engine-out scenarios. The forward wing would be derived from the V-22 while the aft wing, of greater span, employed a root stub with the V-22 structure outboard. Bell estimated that the QTR would be approximately 50-60% common with the V-22, greatly reducing development costs and time.

About the size of a stretched C-130 save for a wider interior, the QTR was to have six times the internal volume of the V-22, twice the vertical lift capability, and accommodate more than 90 troops. A ferry range of 1,749nm at a cruise airspeed of 280-300kts, hover GW of 100,000 lb (45,359kg) and maximum GW of 140,000 lb (63,503kg), were estimated. It was expected that 15,000-25,000 lb (6,804-11,240kg) of cargo could be delivered vertically, or 40,000 lb (18,145kg) STOL, at a range of 540nm.

The utility of a machine like the QTR in sea replenishment and battlefield mobility could be revolutionary, and it attracted the attention of the US armed services. The aircraft could carry such loads as three 'Humvee' vehicles, a Marine Light Armored Vehicle, a UH-60 Black Hawk helicopter, or eight 436L logistics pallets, all loaded via an aft ramp. Under its Ship-to-Objective Maneuver doctrine the Marines sought to move voluminous supplies as much as 200 miles (124km) inland without a build-up on the beach. Likewise, airlift of supplies to a forward airstrip by C-130 or C-17 for breakdown and flight aboard helicopters to forward-deployed troops could be eliminated. The QTR would support the US move toward a lighter and more mobile army. A valid criticism was that it was unlikely the QTR could be designed to fit on the elevators and in the hangar decks of the Navy's amphibious assault ships and carriers. Bell Boeing conceived a gunship version of the QTR as well as commercial derivatives.

Initial discussion suggested a Hercules-size Army FTR (renamed again the Air Maneuver Transport, AMT) capable of carrying 18,000-20,000 lb (8,165-9,070kg) of cargo vertically and at 300kts to 270-540nm. The QTR as originally envisioned would not meet these requirements. It was proposed that a V-22-derivative QTR be introduced and then progressively uprated, earning the unofficial name 'V-44'. Water tunnel testing had already been performed to visualize the complex flow patterns from the tandem wings and proprotor wash interactions. The government provided Bell with several study contracts beginning in 2000. The first was to look at the technology required to convert the V-22 into a QTR. This would be capable of transporting 20-metric tons of cargo or 122 troops. A second funded a hover download wind tunnel test model. One discovery was a 20% reduction in fountain recirculation download through the rotors for OGE hover, and a positive lift effect IGE. An unmanned subscale demonstrator was suggested, with initial consideration given to employing elements of the Eagle Eye. However, the ability of the DoD to pay for development and procurement of a new aircraft during the first decade or two of the new millennium was doubtful. AMT was tenuous and, even as another joint program, yielded fielding dates of 2015-2020. Bell and

Right: **Cockpit and instrument panel of the first Model 609. Flat panel displays have become the standard accouterments for advanced aircraft of this type.** Jay Miller

Below: **The culmination of decades of dreams and hard labor was seen in the first flight of the BA 609 commercial tiltrotor at Bell's Plant 6. At only nine seats, the 609 was a cautious entry of the tiltrotor into the civilian field, but closely tied to the fortunes of the V-22.** Jay Miller

Agusta also conceived a QTR version of the BA 609 as the 34,000-lb (15,420-kg), 26-seat BA 626.

Research into deriving low radar signature and speeds up to 500kts from the tiltrotor focused on swept blades and folding proprotors. The latter could require a 'convertible engine', serving both as a turboshaft power plant for the proprotor transmission and as a turbofan for high-speed cruise. A practical engine of this type has proven elusive. Alternatively, engines could drive the proprotors and 'cruise fans' through shafting and clutches for up to 450kts airspeed. Bell tested a 25-ft (7.6-m) diameter folding proprotor in a wind tunnel during 1972, with the folding performed successfully at up to 200kts. Separate Bell and Boeing design studies, with some USAF funding in 1969 for a 60,000 to 67,000-lb (27,212 to 30,391-kg) Advanced Rescue Aircraft, generated 'tilt-fold' or 'stop-fold' concepts. Bell's D-270 configuration featured four turbofans in pods under the wing, while the D-271 and D-272 were conceived as demonstrators. In late 2000 the company vied for NASA funding to convert an Eagle Eye as a testbed for high-speed tiltrotor research.

Others sought to enter the tiltrotor field. By 1999 Sikorsky was studying the Variable Diameter Tiltrotor. This envisioned a proprotor that could expand to greater diameter during hover flight, even extending over the cabin, and then decrease to about 70% diameter during conversion for higher speed cruise or for landing in full airplane mode. Telescoping and reelable rotors were conceived and some scaled wind tunnel testing performed successfully. The larger diameter would make for improved hover performance by virtue of lower disk loading, increasing single-engine and autorotation capability, lower the groundwash intensity, and possibly realize a noise reduction. Smaller diameter in cruise would reduce gust sensitivity and might not require an rpm reduction. In Europe plans were set in motion as early as 1986 – building on the Aérospatiale X910 design work of the 1970s – to begin 'second-generation tiltrotor' research with an eye toward possibly producing a 12-30-seat commercial machine with a takeoff weight of 22,000-28,600 lb (9,979-12,973kg). This fought for funding, with low-level engineering efforts, for over more than 15 years under EUROFAR, EUROTILT, and 2GETHER titles. The program finally kicked-off in 2000 with substantial research funding through 2005. Agusta and Bell were soon discussing a competing 28,000-lb (12,700-kg), 19-seat BA 619. AgustaWestland's 19-22 passenger, 22,000-lb (10,000-kg), ERICA concept featured the innovation of separately tilting both the proprotors and most of the wing underlying the disks, either together or separately, to get the best of tiltrotor and tiltwing approaches. This would substantially reduce wing download for a payload benefit, and allow a smaller diameter proprotor that can tilt level on the ground, although introducing additional aeroelasticity concerns. In Russia the Mi-30 was conceived variously as an aircraft in the class of BA 609 or V-22, but the program never developed. Yakovlev conceived a 990-lb (450-kg) tiltrotor UAV called the Albatross that was also stillborn.

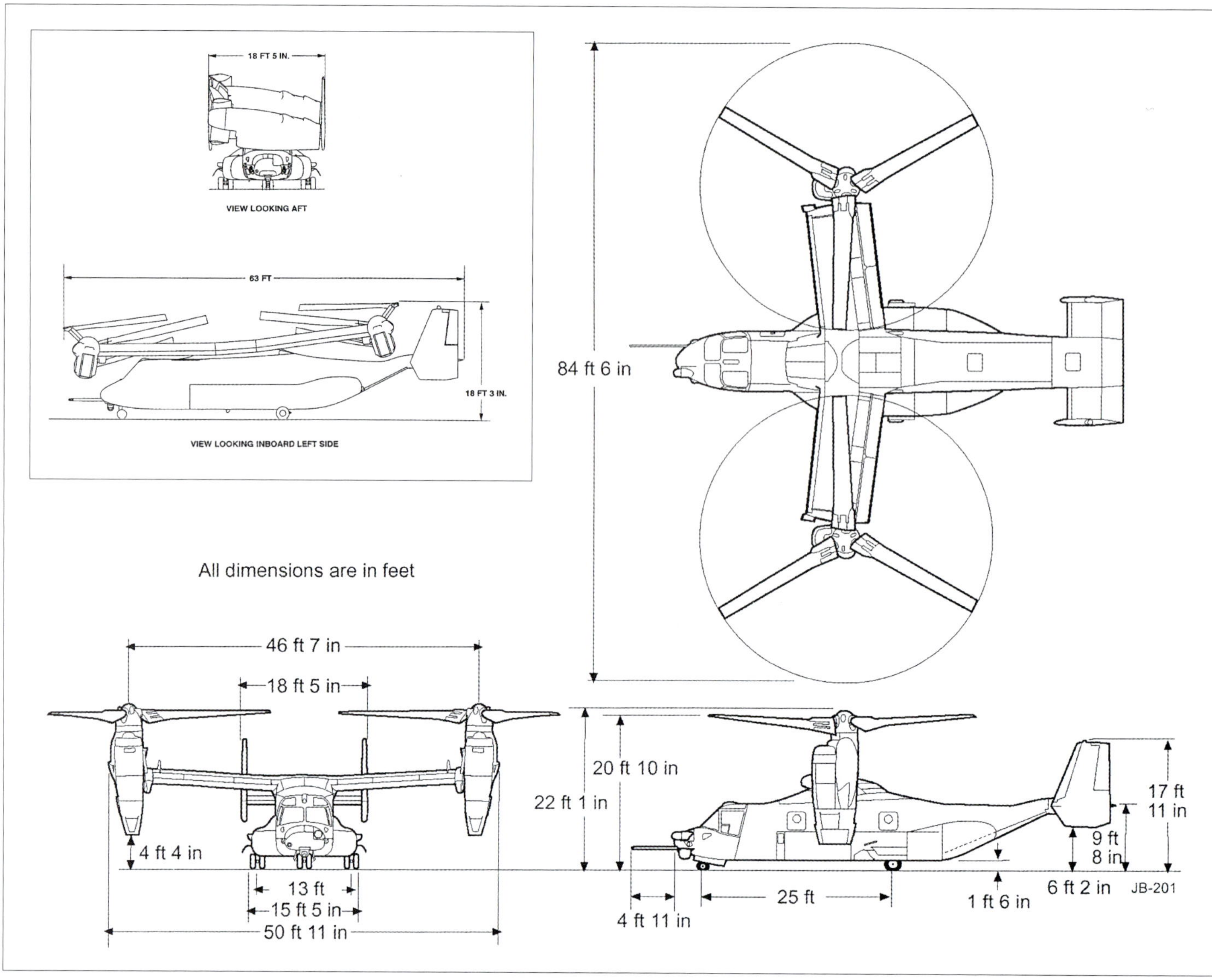
18 FT 5 IN.
VIEW LOOKING AFT
63 FT
18 FT 3 IN.
VIEW LOOKING INBOARD LEFT SIDE
84 ft 6 in
All dimensions are in feet
46 ft 7 in
18 ft 5 in
4 ft 4 in
13 ft
15 ft 5 in
50 ft 11 in
22 ft 1 in
20 ft 10 in
17 ft
11 in
9 ft
8 in
6 ft 2 in
1 ft 6 in
25 ft
4 ft 11 in
JB-201

Bell Boeing

RESCUE
REMOVE BEFORE FLIGHT

Chapter Seven

Osprey Described

The following detailed description represents the aircraft in mid-2003. Changes since that time and other models of the V-22 may render the description inaccurate in some respects.

Structure

The V-22 structure is a semimonocoque design consisting mostly of carbon/epoxy composite material in a frame/longeron construction with stiffened skins. Approximately 50% of the structure is made up of this material, about 10% of glass fiber composite, and some 20% metals. The requirements for a fail-safe structure with high ballistic survivability led to a multi-element redundant design. The engine nacelles are made mostly of titanium (for heat resistance) and aluminum with some graphite epoxy composite. The proprotors are fiberglass/epoxy composite material blades spars and skins with Nomex honeycomb afterbody shapes, carbon/epoxy composite-material blade grips, fiberglass/epoxy rotor yokes, and steel rotor masts.

Preceding page top: **The three-view includes the early short aerial refueling probe, later revised; inset: the 'folded' dimensions are shown.** Both Author's collection

Preceding page bottom left: **Some of the elements standard to the MV-22B and CV-22B are evident in this image, including the twin pitot static probes, angle-of-attack vane, knee window, and windshield wipers.** Author's collection

Preceding page bottom right: **The object protruding just ahead of the side emergency exit is the laser detector set sensor array.** Author

Below left: **Base of the aft fuselage and tail. The aircraft model and Bureau Number are stenciled on the aft end of the fuselage side as seen here on aircraft 9.** Author

Below right: **Details of the fuselage strake.** Author

The wing features a 100-in (2.5-m) chord and 23-in (0.6-m) thick airfoil at its deepest dimension. The thick section helps in low-speed lift generation, and promotes increased wing stiffness to combat aeroelastic instabilities, while also providing greater internal volume for fuel cells. A 6° forward sweep ensures sufficient clearance at maximum proprotor blade flapping. The 3.5° dihedral is principally to ensure fuselage clearance during the blade fold/wing stow operation. The wing is mounted to a 91-inch (2.3-m) diameter steel flexring or 'carousel' atop the fuselage. This permits the wing to be rotated 90° on the ground – parallel to the fuselage – while also transferring flight loads to the fuselage. The wing is locked into the flight position by four pins that are hydraulically retracted before the stowing action begins. More than 1,900 electrical connections as well as fluid lines pass from the fuselage into the wing through this interface without crimping.

A Structural Load Limiting feature of the flight control system helps in controlling proprotor and aerodynamic loads imparted to the structure via maneuvering. Both static load margins and oscillatory fatigue limits are ensured by restricting or modifying the pilot input, or automatically preventing adverse vehicle response. This also assists in preventing unintentional overstress of the airframe by limiting the normal load factor a pilot can command depending upon many parameters.

The V-22 incorporates a number of passive means to reduce vibration throughout the aircraft. Rotor pendulum absorbers consist of three steel arms extending radially from the rotor mast, each with a pendulum rod and 30-40 lb (14-18kg) weight. The swinging frequency of the pendulum is tuned to naturally counter and damp the primary rotation-induced vibration of the rotor; the three-per-revolution (3/rev) excitation. The fence atop the outboard wing surfaces lessens the inboard curving of the vortices shed from the wingtip/nacelle interface that impact the empennage. This reduces random vibration from tail buffeting during conversion. A strake is attached to both sides of the fuselage just forward of the wing leading edge roots. They redirect airflow to reduce the energy of vortices shed from this region that would otherwise impart unacceptable buffeting loads to the vertical tails.

The Active Vibration Suppression System consists of masses in the cockpit area driven by electromechanical motors at frequencies determined by a computer using structural response measured at the cockpit. This reduces or cancels the dominant 3/rev vibration mode by producing an equal and opposite force. Up to 1,000 lb (436kg) of force can be generated, with only the forward fuselage directly benefiting.

The V-22 is designed for high occupant survivability in the event of a crash landing. The cockpit is designed to allow impacts up to 25G to be survivable. The aircraft can be belly landed as an airplane to reduce the high sink-rate impacts common with helicopter autorotation landings. The main nose bulkhead is installed with the top angled forward. This anti-plow feature helps to protect against the aircraft overturning during crash-landing with forward momentum. The graphite-fiberglass proprotors are designed to shred on contact with the ground under power, eliminating the hazard of high velocity material. Also, the direction of rotation should send most material away from the fuselage. Engine nacelles and even the wing can depart the aircraft with little risk of fatal damage to the fuselage. In the event of a water

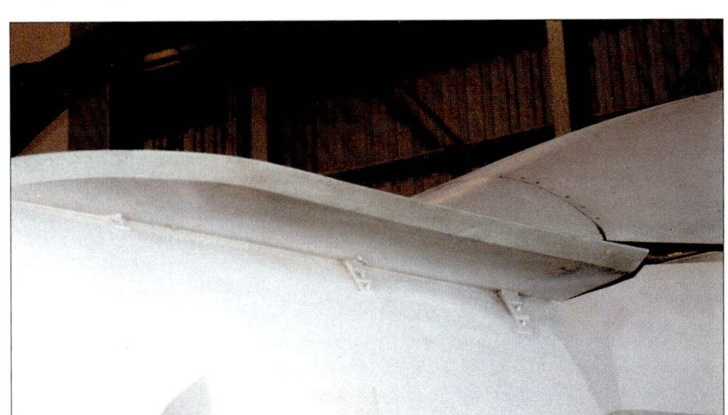

Top left: **Crew entry/exit door.** Jay Miller

Top right: **Four static discharge wicks are installed at the trailing edge of the vertical tail, with three on the rudder. Note the color demarcation at the base of the surface. Initial installation of the SIRFC transmit antenna was at the forward end of the tail.** Author

Bottom left: **A view of the partially closed cargo ramp and door, with pads at the end of the ramp and an access panel in the door.** Author

Bottom right: **Emergency egress paths include those from exits that must be opened by pyrotechnic charges. As many as 28 persons must be able to evacuate the interior within moments.** Bell Helicopter

landing, the wide base of the V-22, its voluminous sponsons, and the tip engine nacelles should provide good buoyancy following a controlled ditching to prevent the rollover common with helicopters. It should remain upright for up to two hours to allow safe egress.

Three external aircraft tiedown rings incorporated into each sponson outer profile allow the aircraft to be quickly chained down to a vessel's deck or shackled to an apron ashore. These are sufficiently strong to restrain the aircraft in 100-kts winds. Jacking points are provided on the fuselage body and main landing gear struts. The V-22 also has hoist points for crane lifting or aerial recovery. Pull-down footsteps are also found in the sponsons, plus the fuselage side and aerodynamic fairing at the wing junction to the fuselage incorporates steps and handholds, for personnel to climb atop the aircraft. The top of the fuselage, sponsons, wing and horizontal tail have strips of no-skid material to allow personnel to move about safely without risk of slipping when these surfaces are wet. The no-skid also serves to denote those areas where treading on the structure is permitted. The wing upper surface has points where poles can be inserted to which personnel safety harness tethers can be attached, preventing falls.

Blade Fold/Wing Stow

The BFWS reduces the dimensions of the V-22 for shipboard deck and elevators movement, and stowage on hangar decks. The BFWS button in the cockpit overhead console is held depressed while the action is performed automatically, powered by the auxiliary power unit (APU) or alternatively by a ground power cart. The action can be interrupted and resumed at any point.

The BFWS commences with the nacelles rotated to the 90° 'palm tree' position and flaperons raised. The proprotors are turned to 'index' the one non-folding blade of each proprotor over the wing. The four other blades are then folded parallel to the indexed blade via an electric motor within each spinner. The pins locking the wing in the flight position are retracted hydraulically. The nacelles then begin rotation to 0° while the wing rotates clockwise until parallel to the fuselage. A hydraulically operated drum rotates to take-up and play-out a steel cable wrapped around the ring on the bottom of the wing riding upon the fuselage 'carousel' to affect the rotation. Once the rotation is completed a single locking pin is engaged to secure the wing. The BFWS operation is automatically halted should any component position not correspond with proper and safe orientation. Returning the wing and blades to a flight position is essentially the reverse of that for stowing.

The entire BFWS operation takes about 90 seconds and can be performed in a 45-kt wind on a pitching and rolling deck. The procedure can also be done manually in 10 minutes. Nearly all maintenance can be performed with the aircraft folded.

Cockpit

Following helicopter convention, the pilot position is on the right side of the cockpit and the copilot on the left. A cyclic stick is provided in front of each pilot. A pair of throttle-like Thrust Control Levers operates both engines simultaneously. The TCL for the pilot is installed on the starboard side of the center console while the copilot TCL on the left side console. The cyclic and TCL grips contain a number of switches to reduce the need for pilots to remove their hands from the controls. Standard yaw pedals are provided. These also command nose wheel steering and are fitted with toe brakes. Adjustable seat height and rudder pedal distance accommodates a broad population of pilot dimensions.

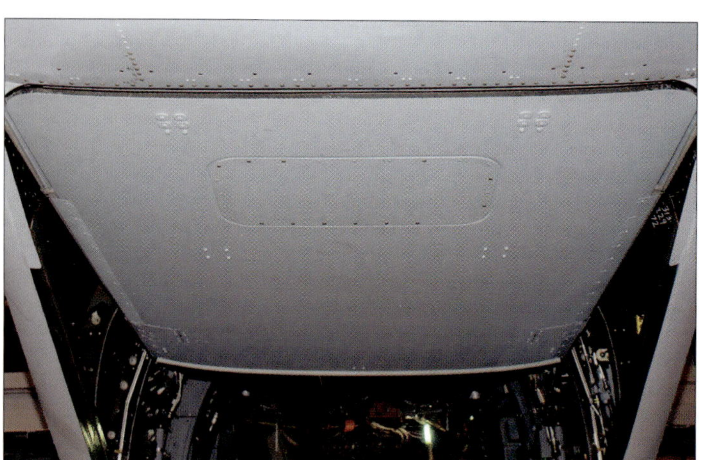

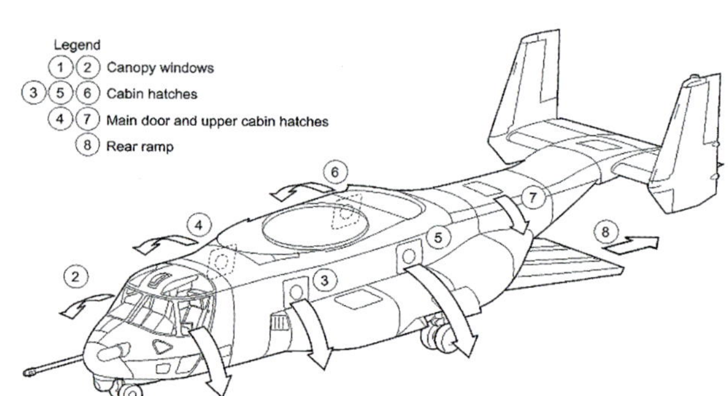

Emergency Exit Locations

The Cockpit Management System replaces the conventional cockpit layout of individual instruments and control heads as the primary crew interface with aircraft systems. This is primarily done through four color multi-function displays that dominate the instrument panel. They provide extensive flexibility in display composition to suit mission and situation. A Standby Flight Display in the center of the control panel is supplemented by conventional instruments providing essential information in the event of a failure rendering the MFDs inoperative. A glare shield that holds the Flight Director Panel, plus other controls and indicators, tops the instrument panel. It also houses, in front of each pilot, Remote Frequency Indicator Selectors, and switch/indicators for MASTER ALERT and PFCS FAIL/RESET. The Control Display Unit/Engine, Instruments, Crew Alerting System (CDU/EICAS) display and twin keyboards are located at the forward end of the center console.

The MFDs are individually programmable to provide flight symbology, sensor video, communications-navigation, and system status displays. Switches control display intensity and contrast, and brightness for day or night operations. Nineteen bezel keys surrounding each display allow selection of functions or modes identified by adjacent legends on individual 'layers' ('pages') of displays. The MFD moding cursor control on each cyclic grip permits pilots to select the mode for the switches, and which MFD is being addressed. There are five principal flight display modes or 'top layers'. Flight displays include the Primary Flight Display (with or without FLIR image underlay), Vertical Situation Display (with or without FLIR underlay and control layer), and hover/transitional flight display. Navigation functions includes the Horizontal Situation Display (HSD, with or without moving map underlay), HSD control functions, flight plan sequencing, navigation (nav) aid selection, nav update and store, nav system setup, and Digital Map System (DMS) controls. The FLIR functions include FLIR display, FLIR controls, and FLIR setup layers. Status functions provide graphical status pages of aircraft system functions and textural status pages. Finally, system functions displays include operator functions for the display (such as declutter) and maintainer system functions.

The CDU/EICAS display and two associated keypad units dominate the center console between the pilots. This is primarily used to dis-

Above: **Some of the blade fold mechanism is evident as well as the proximity of the left hand nacelle to the nose.** Author's collection

Below: **The MV-22 cockpit reflects modern design practices. Photographed from just inside the cockpit door, the space aft of the center console would be occupied by a crew chief in the jump seat folding down from the door.** Boeing

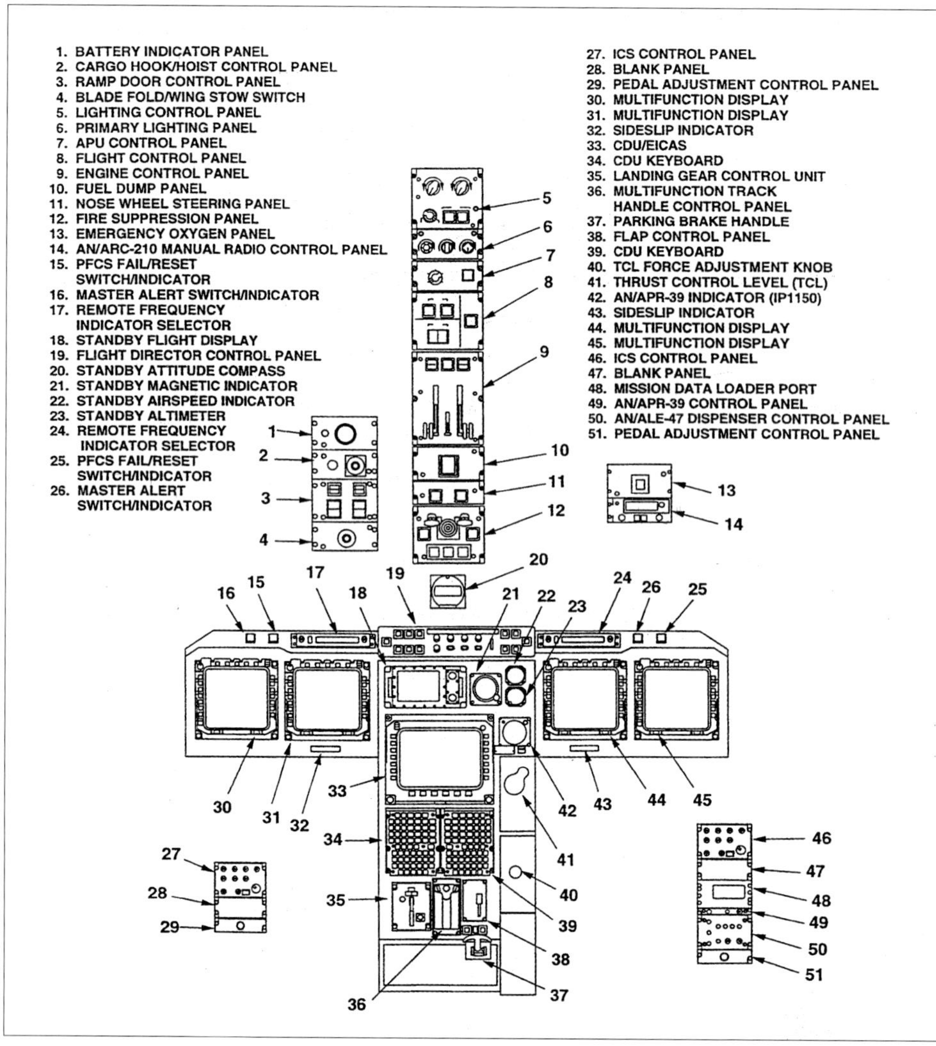

Left: **The generic MV-22 cockpit instrument panel shown with side and overhead consoles. The CV-22 varies somewhat and years of service will see other changes.** Bell Helicopter

Bottom left: **The cyclic grip with imbedded switches.** Bell Helicopter

Bottom right: **The switches and controls for the Digital Map associated with the Track Handle are detailed here.** Bell Helicopter

play cautions and warnings, and provides graphical depictions of systems for control and to diagnose faults. It functions much the same as the primary MFDs. The keyboards allow rapid access to the radio and navigation setup, Identification Friend or Foe (IFF) control, electronic flight plan management, mission summaries such as fuel and cargo, emission control to inhibit radio transmission, plus windshield wiper, environmental control system (ECS), and exterior lighting control. The console also contains a radar warning system display, the landing gear lever, flap handle, and a track handle. This handle slews the FLIR, digital map display, and radar, and has provisions for operating a nose gun.

Cockpit lighting is compatible with night vision goggles. The NVG is the AN/AVS-6 Aviator Night Vision Imaging System, commonly called 'ANVIS 6'. This includes a HUD Display Unit (DU) monocle that can be fitted on either side of the goggles. The NVG/HUD shows flight and navigation data in various display formats allowing the pilot to fly the aircraft at night with minimum reference to cockpit instruments. The DU cable connects to the vehicle and control is via a unit in the overhead console, with display layer selection also via the Helmet-Mounted Display switch on each cyclic.

The cockpit is fitted with large windscreens, side windows, overhead windows, and knee windows. The windscreen is designed for birdstrike protection and the entire cockpit provides aircrew protection from rifle-caliber rounds. A rearview mirror is fitted on the windscreen bow at the forward end of each side window. The side windows can be jettisoned for emergency egress. Jettison actuators are on the forward windshield posts as well as external, firing pyrotechnic charges. The armored pilot seats incorporate a vertical stroking feature to attenuate crash energy and reduce the likelihood of injury.

The cockpit door opens into the cockpit passageway. A jump seat installed on the cockpit door can be folded down after closing the door. An evaluation pilot or USMC crew chief may occupy the seat. But, it is essential seating for the USAF flight engineer in the CV-22 to assist pilots in operating the extra equipment supporting their mission.

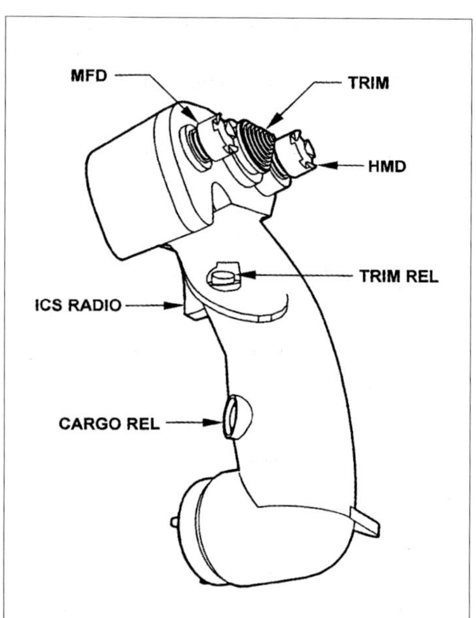

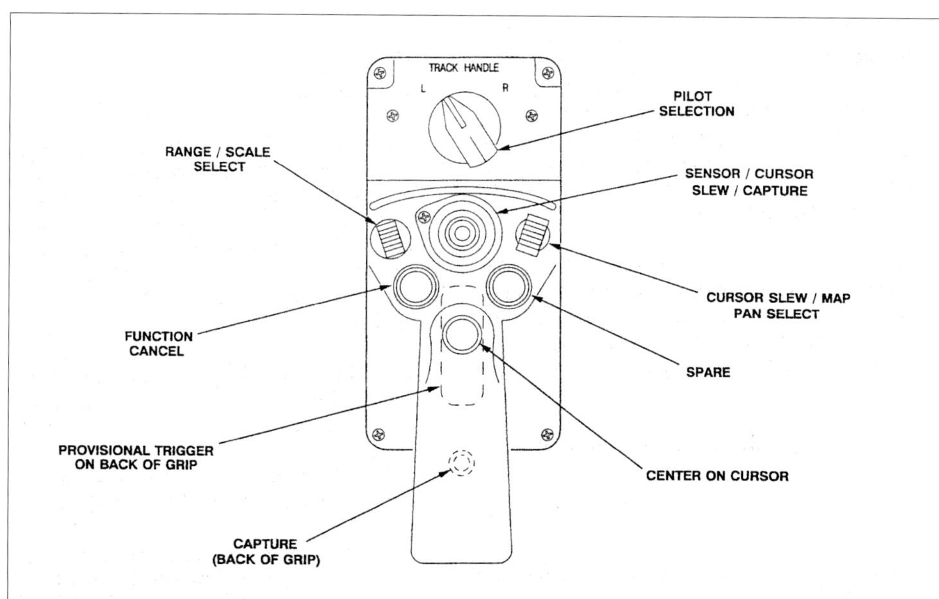

Cabin

The cabin is divided into four zones, each permitted 8,850-lb (4,014-kg) maximum cargo weight. The floor is made up of seven Nomex honeycomb panels covered with graphite epoxy and fitted with 40 cargo tie-down rings in four rows. Four tie-down rings are also located in two fuselage frames and four on the cargo ramp. The two types of tie downs are rated to 5,000 lb (2.268kg) and 10,000 lb (4,536kg). Thirty-five receptacles for mounting raised cargo roller rails are also installed in the floor, and 15 on the ramp. When not in use the rails are stowed at the base of the cabin sidewall. Aluminum shoring is available to help distribute tire point loads of wheeled vehicle. These planks have holes placed to allow access to tiedown rings.

A fold-down, side facing, flight engineer seat is installed on the left avionics cabinet at the forward end of the cargo compartment. Twenty-four plastic fold-down troop seats, each crashworthy/energy attenuating to 6G, are mounted to the cabin sidewalls.

Three side window/escape hatches and an overhead ditching hatch facilitate emergency cabin egress. Jettison actuators activate pyrotechnic charges to open these hatches. The 3-ft (0.9-m) wide main cabin door in the starboard forward portion of the cabin is the primary means for crew entrance and exit. The upper portion slides up on tracks or can be removed for emergency egress. The bottom of the door folds out and incorporates steps. Because the door is in the path of the proprotor in APLN, the door is automatically prevented from being opened while in flight with the engine nacelles between 0° and 45°. The door can be opened in VSTOL or CONV for the purpose of scanning outside the aircraft and hoisting someone aboard. If the door is open, safety interlocks prevent the engine nacelles from being rotated forward beyond 45°. In VSTOL, on or close to the ground, a good deal of airflow enters through the door, especially if the ramp is down. This can carry ground material lifted up by the groundwash into the cabin. One job of the crew chief or flight engineer is to sweep out the cabin.

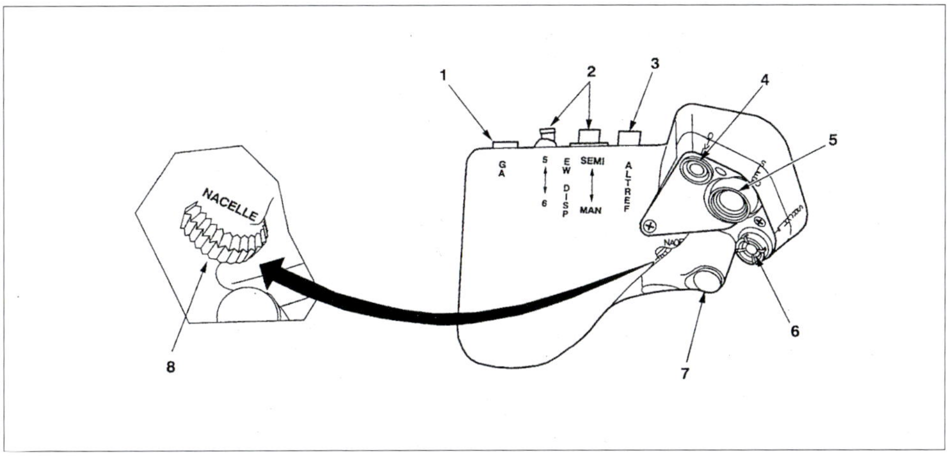

Top: **The broad thrust control lever grip is adorned with many switches. Detailed are: the Go Around (1); CMDS dispense (2); Altitude Reference (3); TCL Overtravel (4); FLIR Slew/Capture (5); Searchlight control (6); provisional (7); and Nacelle Control thumbwheel switches.** Bell Helicopter

Middle: **Forward tilt of nacelles provides VSTOL performance capability**. Bell Helicopter

Right: **The cabin is shown from the ramp sill to the cockpit door with the troop seats secured up with velcro straps. Note the cargo roller trays stowed at the base of the sidewalls under the folded down buffer boards and the life raft stowage in the forward right corner.** Author's collection

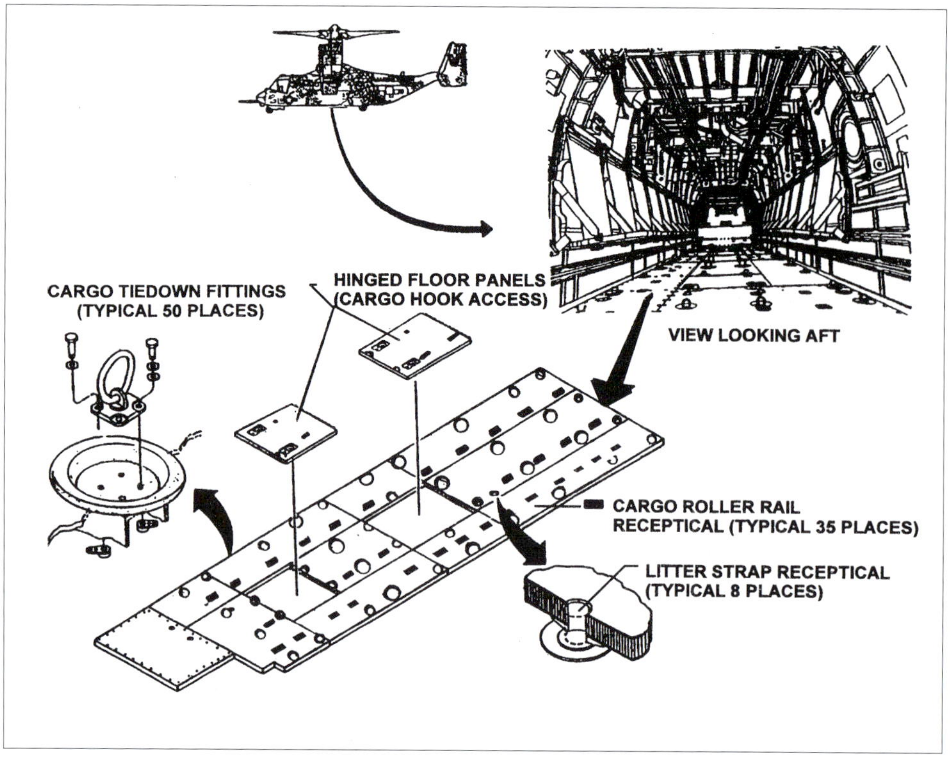

Left: **The cabin floor layout includes the removable panels to access the cargo hooks, receptacles for cargo roller trays and litter stanchions, and the 50 cargo tiedown rings.** Bell Helicopter

The cargo door rotates inwards and up against the ceiling while the ramp is lowered outward and down for troop and vehicle onload/offload. The ramp and door are each operated by two actuators. They are controlled and monitored via a panel just aft of the main cabin door, the panel in the cockpit overhead console, and an externally panel in the aft starboard sponson. They may be operated when aircraft systems are unpowered by using a battery driven hydraulic pump. The ramp can also be unlocked and lowered manually via a handle on the ramp control valve mounted in the aft cabin.

Two ramp extensions can be installed at the end of the ramp in many different positions to accommodate various types of roll-on/roll-off cargo. The extensions fold back flat against the ramp for stowage or are removed. When palletized cargo is loaded from a wheeled loader, the ramp is only lowered to the horizontal position. This configuration is also used for troop or cargo airdrop. A rise in the aft fuselage just forward of the tail assembly permits cargo to clear the structure during roll-on and roll-off via the ramp.

The cabin contains everything needed to load, secure, and off-load cargo and troops. A barrier net and cargo stop can be mounted in the cabin to ensure the cargo does not shift forward. Buffer boards mounted along the lower portion of the sidewalls help prevent the cargo from striking the airframe. A 2,000-lb (907-kg) capacity variable-speed hydraulic winch in the forward cabin, and a pulley system, allows heavy cargo to be pulled into the aircraft via a 150-ft (46-m) cable and hook. A forward cabin panel includes a switch to cut the winch cable via guillotine in an emergency. The primary control is the Hoist/Winch Operator's Grip (H/WOG) stowed just below this panel. It is on the end of a 16-ft (5-m) cord, connected either at the forward or aft end of the cabin, permitting mobility throughout the cabin and even outside. The Block 10 CV-22 was to have hoist controls in the cockpit.

Cargo barrier, winch, and tiedown gear is stowed in the cabin sidewalls. Emergency equipment in the cabin includes fire extinguishers, first aid kit, two 14-man life rafts, life jackets, and an underwater acoustic beacon.

Two interior floor panels and external doors open for access to the forward and aft external cargo hooks. External loads are carried using inverted-V and inverted-Y cable configurations. Using both hooks helps to stabilize the load and permits higher speeds. In the event that external loads become unstable or some other emergency condition ensues, commanded or automatic external load jettison is available. The command can be given from the cockpit or the cabin controls for electrical hook release. The automatic option arms both hooks for release in the event either hook senses a loss of loading. Manual hook release is via a lever located in the area of the aft cargo hook.

The hoist in the cabin door provides the means of lifting personnel into the aircraft. The lower portion of the cabin door is lowered to a vertical position to allow a person to pass unobstructed. The hoist is mounted to the forward cabin bulkhead adjacent to the crew door. It is rotated up and out, locking into position. The hoist has 245ft (75m) of 5/32-inch (4-mm) steel cable. There is 20 inches (0.51m) of clearance between the cable and the aircraft skin. A cable guillotine is fitted in the hoist for emergency use. The hoist is operable from the cabin as well as the cockpit.

An interesting feature of the H/WOG is remote hover trim control, used in conjunction with a two-axis thumb controlled joystick adjacent to the crew door. These controls permit limited lateral and longitudinal control of the aircraft by the operator of the hoist in the cabin door during rescue operations, precisely positioning the aircraft over the desired pickup point.

Personnel airdrop is either freefall off the ramp or static line jumps. The static line installation was still under development at time of writing, but was likely to be a single line for two 'sticks' of six jumps going off either side of the ramp. Troops may rappel, or fastrope, from the ramp or, alternatively, from the cabin door. This involves lowering of one or two ropes from the hovering aircraft to allow troops to rapidly slide

Below: **The raised floor panel and open bottom doors reveals the aft cargo hook, rotated to the deployed orientation, in FSD testbed aircraft 3. The electrical connections are for the load jettison hook release function.** Author

down to the ground in areas where landing is impractical. From the horizontally positioned ramp, the 2-inch (50-mm) Fast Rope Infiltration/Exfiltration System lines are attached to a deployed trapeze with cross beam above the aft end of the cargo ramp. The MV-22 rope has lead woven into it to help prevent it being blown too far aft by downwash. Fastrope exfiltration consists of as many as six persons on each of the two ropes (only two on a forward rope) holding on as the aircraft flies away. This allows a rapid departure if immediate landing to onload the troops is impractical. An alternative is the Special Insertion and Extraction rig or Stabilized Extraction system that has clips along the rope so that the soldiers can hang from the rope with their hands free to fire weapons. The MV-22B is also expected to continue the mission of deploying troops into water, most commonly by 'helocasting' a zodiac craft into the water from the ramp followed by the swimmers.

Flight Controls

The Primary Flight Control System (PFCS) and the Automatic Flight Control System (AFCS) make up the Vehicle Management System in a triple-redundant digital fly-by-wire system. This means that there are three independent flight control computers (FCC) with only one required for flight – although a mission would normally be terminated in the event of a FCC failure. The computers command the surfaces, nacelle angle, and proprotors hubs using three hydraulic systems. System components are separated to the extent possible to increase survivability such that one hit by an enemy round will not prove catastrophic. Each channel is monitored for catastrophic faults and dropped off-line should one be detected. Each FCC communicates with other systems via dedicated data buses. Airspeed and altitude are sensed via four heated pitot-static probes mounted in pairs on either side of the nose. Altitude sensing is supplemented by a RADALT good to 4,500ft (1,370m).

Pilot control movements are translated into electrical signals by transducers and fed to the FCCs. The system adjusts the input and the feedback, with flight behavior tailored and vehicle stability augmented. Hence, it automatically operates some aspects of flight control without the need for pilot intervention to make the aircraft easier to fly. The AFCS is the principal source of these functions, although carried out by the PFCS. It provides automatic control and maneuver adjustment for such aspects as control authority and rate limiting, plus stall and over-g protection. It also provides unique control modes for specific phases of flight such as the typical attitude hold, heading hold, and turn coordination. For example, the system automatically compensates for asymmetric proprotor thrust and torque splits that occur with roll rate, plus dynamic maneuver artificial damping, among many other beneficial features. An important characteristic is conversion protection that automatically rotates the nacelles or modulates the tilt rate to correspond with the allowable conversion corridor airspeed limits. This also prevents wing stall or excessive rotor oscillatory loads and empennage buffet that can promote fatigue damage while the pilot seeks the best performance for the circumstances. If the high end of the corridor is exceeded, the AFCS will rapidly reduce the nacelle angle with a resulting increase in airspeed and loss of altitude if the pilot fails to compensate. However, it prevents sinking during conversion from VSTOL to APLN by reducing the nacelle actuation rate. The pilot can override conversion protection nacelle motion with a full nacelle movement commanded via the thumb switch.

The V-22 is statically and dynamically stable and could conceivable be flown without a FBW system. However, handling qualities would be poor in some regions of the envelope and require considerable pilot compensation at a high workload. Many operational limitations would have to be imposed to replace the automatic protection aspects of the electronic flight controls.

The fundamental means of controlling the tiltrotor was explained in Chapter One, although the V-22 application has many electronic aids making control of the aircraft not entirely intuitive. The automatic phased mixing

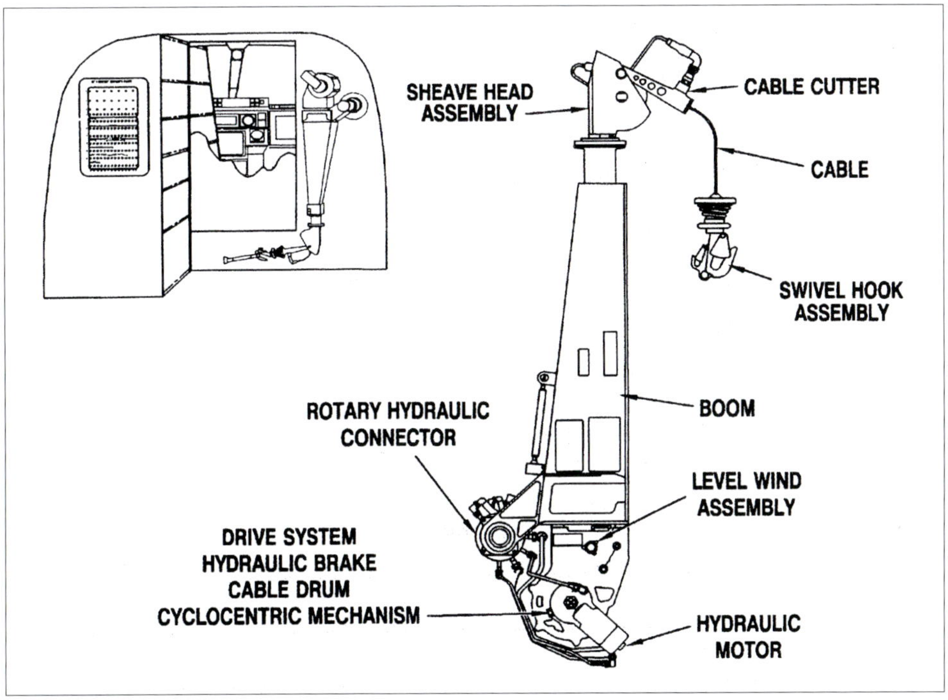

Top: **Details of one of the two external cargo hooks on an EMD aircraft.** Author

Middle: **As troop mount up, aft fuselage details are observed. These include the toes folded back against the ramp.** DoD

Bottom: **The rescue hoist design and installation was still subject to change at time of writing, but was expected to appear as seen in this drawing.** Author's collection

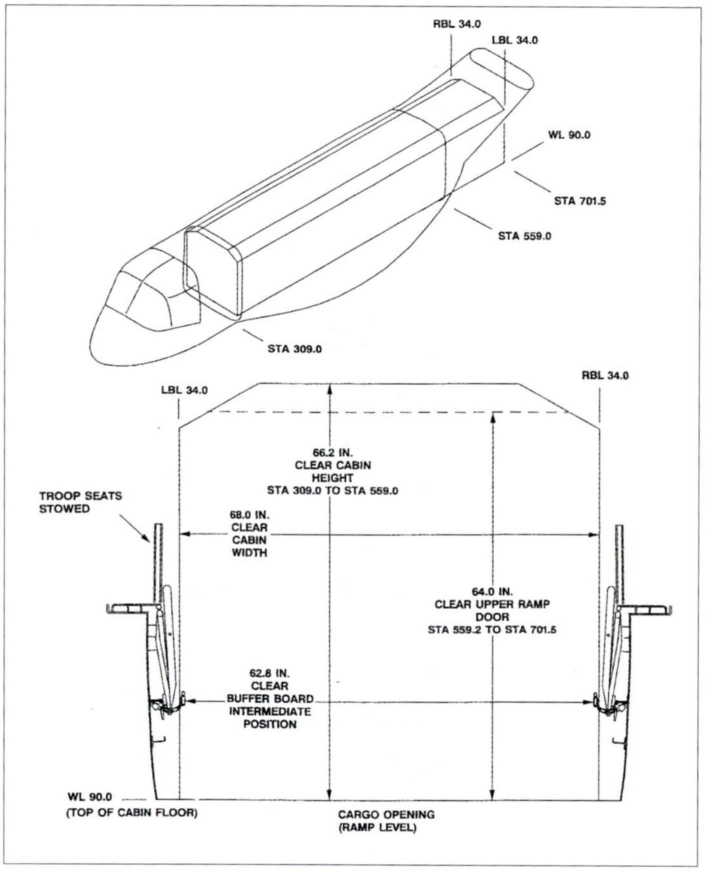

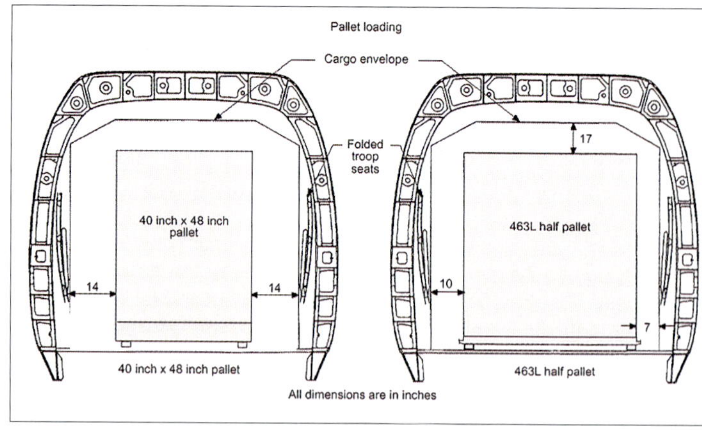

of airplane and helicopter control and the requirement to counter undesirable dynamics or aerodynamics makes the earlier explanation somewhat superficial. For example, the proprotor collective pitch and engine power is automatically modulated to satisfy the TCL torque command and maintain rotor rpm while providing over-torque protection.

Proprotor cyclic and collective control is affected via three actuators displacing a fixed swashplate, that in turn displacing a rotating swashplate that changes the blade angles to tilt the proprotor disk or change blade pitch. Within the mast is a slipring that transmits power and dozens of electrical signals for proprotor functions through the rotating interface. The FCCs also signal the engine FADEC units to alter fuel flow for RPM adjustment. The amount of power increase with forward TCL movement is determined by the FADECs based on the mode of flight, nacelle angle, and other factors. Proprotor pitch is controlled automatically in airplane mode to maintain a constant rpm and reduce blade flapping. Helicopter and airplane control logic, through the cyclic and TCL, are smoothly faded together during conversion, with 60° nacelle angle the point of approximately half of each. The flaps are used in both VSTOL and APLN, but roll control with flaperons, rudders movement and elevators are locked out in VSTOL. Full elevator authority is available up to 140kts.

Control surfaces on the H-shaped empennage consist of a single-piece elevator and two identical rudders. The elevator, being a critical surface, is powered by three hydraulic actuators, each driven by one of the three independent hydraulic systems. Each rudder is driven by a single actuator and different hydraulic systems, and always move identically. Failures allowing the use of only one rudder would not be severe as rudders are not critical in most circumstances, and one alone is usually sufficient. A horn extending ahead of the rudder hinge line at the top contains mass for dynamic balance to ensure stability in unpowered conditions.

Full-span flaperons are split into two segments per side for safety redundancy, although the twin segments always moving together. Moving down as flaps the surfaces move aft to open a slot between it and the wing. An autoflap feature will automatically command symmetrical flap settings appropriate for the mode of flight and airspeed. Autoflaps uses high settings up to 72.5° below 40kts to reduce wing download. The 40° setting is used between 50 and 160kts to augment wing lift and improve pitch attitude control during transition. Between 160 and 200kts the flaperons are varied between 0° and 40° as a linear variation with airspeed to prevent ballooning during reconversion. The pilot may manually select flap settings of 0°, 20°, 40° and full. Differential flaperon movement side-to-side produce rolling moments. The flight control system will ensure that manual flaperon settings beyond 40° cannot be selected with the nacelles below 30° to prevent loss of roll control effectiveness due to flow separation off the flaperons. Each flaperon segment is powered by two hydraulic actuators, each of the eight total being driven off on one of the three hydraulic systems in a scheme that helps to ensure that all segments can be moved in most system failure scenarios.

Top left: **The cargo compartment dimensions are provided in this drawing.** Author's collection

Top right: **Clearances around standard-size palletized cargo is minimal in the V-22. Reaching gear and control in the aft cabin, or to scan outside from that area, becomes essentially impossible with such cargo configurations.** Bell Helicopter

Above: **Installation of the stanchions for the dozen casualty litters leaves just enough room for attendants and the crew chief to move about in the cabin and to access emergency equipment.** Bell Helicopter

The PFCS electronically operates all of the aerodynamic surfaces and rotor actuators, with the appropriate gearing, and phasing in and out of swashplate and aerodynamic surface control, as would a mechanical control system. One important function is sensing longitudinal rotor flapping (blowback) during transient maneuvers and high speed, particularly as the nacelles rotate forward, and protecting the aircraft from high blade loads and mast bumping. This allows the aircraft to operate at the corners of the envelope with less concern.

A single conversion actuator per side, between the nacelles and the wingtips, performs nacelle rotation. These are telescoping hydraulic ballscrew actuators. Because of their criticality, each is operated by the two primary hydraulic systems, with the auxiliary system as a backup. A backup means for moving the ballscrew via an electrical motor in the FSD aircraft was eliminated in EMD. The nacelle moves at 4-8°/sec, proportional to TCL switch displacement. Conversion between VSTOL and APLN requires 12 seconds.

A means of helping the pilot remain within the conversion corridor is an arc of permissible nacelle angles for the existing airspeed displayed on the primary flight display MFD page. In APLN, with the nacelles on the downstop, the rpm limit will prevent exceeding the conversion corridor speed limit. The upper boundary is not constant but varies with Mach number, and the corridor is programmed to be compatible with gross weight. After moving the nacelles to the downstop with the TCL switch, another quick input 'autobeeps' Nr from 100% (390rpm) to 84% (330rpm), or to return to 100% in preparation for reconversion. 'Beeping' is also used to refer to use of the nacelle rotation switch to make fine adjustments to tilt angle.

Helicopter mode refers to flight with the nacelles at between 80° and 97.5°, but commonly 90° (fully vertical). There are mechanical stops at the 0° and 97.5° (7.5° aft) positions to ensure against exceeding these limits. VSTOL mode has been used to refer to flight with the nacelles between 90° and 60°, while conversion mode is with the nacelles at 84° to 1°. Transition has come to refer to flight with the nacelles in motion. Nacelles at 60° are a convenient position where the pilot can transition to airplane flight logic from helicopter logic. Nacelles 0° or horizontal, on the mechanical downstop, is airplane mode.

The Flight Director (FD, the autopilot) core modes are barometric and radar altitude hold/select, airspeed hold/select, heading hold/select, and hover hold. Hover hold keeps the aircraft in a hover, hands-off, with less than 1° heading error, less than 1° of pitch and roll variation, and with no more than 20ft (6m) drift through 2 minutes – steady enough to allow external hoist pickups. The other core modes allow hands-off cruise, departures, and descents. Other FD modes are Inertial Navigation and Electronic Navigation. Coupled navigation modes include Instrument Landing System, Tactical Air Navigation, and Navigation Reference Point. These are guided modes only, with autohover the only 'hands-off' feature. Approach-to-hover and departure-from-hover/go-around modes remained to be perfected.

Flight Characteristics

The V-22 tiltrotor configuration has many unique flight characteristics, some that can be advantageous and others requiring training to employ properly and avoid hazards.

The proprotor wash over the wing delays and masks wing stall while in APLN, with little or no buffet as natural warning. As lift degrades the aircraft begins a 2,000-2,300fpm (10-12mps) descent and the artificial stall warning goes off. If no action is taken the sink increases markedly. An accelerated stall, in a turn, is characterized by wing drop that makes the stall warning essential. Also, the V-22's comparatively small rotor diameter makes it highly tolerant of wind gusts and turbulence.

The rotation of the nacelle forward to APLN produces a forward shift in the aircraft's cg. The reverse is true for reconversion. The nose would tend to follow the nacelle movement, primarily because the thrust line is above the cg and because of the slight cg shift with nacelle motion. This movement is automatically countered by the flight control system to some extent. The pilot frequently must manually trimmed residual pitch, although this may yet be corrected. The cg also moves forward with fuel burn, although not as quickly.

A welcome characteristic of the tiltrotor is the use of nacelle angle for energy management, allowing rapid changes in airspeed and descent rate. Indeed, the V-22 can accelerate from a hover to 230kts in a remarkable 18 seconds, then decelerate and reconvert to a hover in an identical period. CONV gives best angle of climb performance while APLN provides best rate of climb.

The V-22 can operate safely from a 9° upslope, downslope, or cross-slope, with greater capability demonstrated. Up-slope and downslope landings are fairly routine in the V-22 because of the ability to tilt the thrust vector, or keep the proprotor disk vertical and the deck angle parallel to the sloping surface. This also ensures maximum swashplate control authority is available for maneuvers and dealing with gusts or other disturbances. This is especially helpful in an upslope departure where care is required to guard against excessively nose-high attitudes and a tail strike. The V-22 also has generous control and safety in cross-slope landings, requiring as much as 33° of cross-slope angle for a static rollover. However, before a rollover occurs the weight on the downslope main and the nose gear would allow the airplane to slew around into a safer downslope heading.

Osprey low-speed flight close to the ground has idiosyncrasies that require pilot familiarization. One is the relative wind critical azimuths, which are 30-60° and 300-330° with 20-45kts. Maneuver or acceleration into these avoid regions in VTOL mode are characterized by uncommanded pitch-up with sideslip. PU/SS is a consequence of upwind rotor downwash on the horizontal tail during sideslip or quartering headwinds giving a nose-up pitching moment. The pitch-up reduces forward vision, causes rearward drift, and robs the pilot of full longitudinal control margin that may be required to deal with gusts, all increasing workload. A PU/SS compensation is programmed into the control laws that rolls the nacelles forward 2-5°.

Right: **A weighted rope and a team member stabilizing the end of the line allow fairly routine fastrope from the Osprey. Three personnel on the rope at one time also helps prevent the rope being blown aft.** Author's collection

Left: **Looking outboard at the flaperon root fairing on the nacelle and into the wing cover ahead of the control surface.** Author

Bottom: **Looking inboard along the flaperon reveals the hinges and actuator rods.** Author

In another scenario, if the pilot rises to a hover too quickly the aircraft may pitch up more rapidly than the flight control compensation can respond and the pilot runs out of pitch authority to arrest the pitch-up. The loss of control margin could prevent checking aft drift. Pilots are trained to lift to a hover carefully with slow TCL application. Should the AFCS sense full longitudinal control power under such conditions, it will slowly move the nacelles down toward the limit autobeep position of 85° to restore stick margin. The pilot must be aware of this during such operations as IGE hover pedal turns, and shipboard launch and recovery. The best practice is to orient the aircraft's nose toward the windline as soon as it is lifted off the deck or during approach to the ship's deck.

The V-22 tends to pitch down during IGE hover and the beginning of forward motion. This is due to upwash from the ground onto the bottom aft fuselage and tail, requiring pilot anticipation and compensation. Additional power is also required in sideward translation owing to the large side force on the boxy fuselage and the flow from the upwind proprotor blowing across the wing to degrade the performance of its twin. Rearward flight at 10-20kts requires excessive control manipulation. Rearward flight or flight with tail winds is entered with aft nacelles to prevent an unstable pitch divergence.

At full mission weight a vertical takeoff is not possible without a short ground roll to gain translational lift. For some mission gross weights where vertical takeoff is practical, sustained flight with an engine failure is not possible. Consequently, for shipboard operations under these conditions a STO is preferred for a flyaway capability should an engine fail. STO is usually performed with 60° nacelle angle. Demonstrated on the amphibious ships, with moderate GW and wind over the deck, the roll could be as short as one rotation of the main gear tire. At high GW a roll of a few hundred feet may be necessary. The lack of a longitudinal cyclic control position indicator during STO occasionally leads to either 'wheelbarrowing' (nosewheel lift off last in a nose-low attitude), over-rotation (undesired pitch-up after takeoff), or porpoising after liftoff. Careful training helps to avoid these behaviors.

Taxi is normally initiated with 80° nacelle and 18-20% power, speed modulated with nacelle and minimal braking. A normal takeoff begins with a smooth and positive rise to a 20-30-ft (6.10-9.14-m) hover, lifting the nose wheels first. The nacelles are then beeped forward at about 1°/sec, trimming longitudinal stick to compensate, to begin the transition. The gear is retracted and TCL is used to establish the desired climb and airspeed. Once established in APLN, the power is beeped to cruise rpm. For reconversion the pilot slows below 220kts and beeps to 100% Nr in preparation to rolling the nacelles back. The blades at the flatter pitch also create more drag in APLN to help in the deceleration. Nacelle rotation is usually begun at between 150-200kts, depending on GW – higher reconversion airspeed for higher GW. Landing pattern work is typically performed at 60° nacelle and 110kts.

Shipboard departure and recovery are modified versions of that for USMC helicopters. Takeoff involves rising to a hover and translating to port over the sea before accelerating and climb away while converting to APLN. Recovery begins with the aircraft passing the vessel to starboard on its steaming heading in APLN, crossing ahead of the bow, and entering the landing pattern to port while reconverting. The aircraft is swung in and approaches the deck at an angle. Once waved in to land, the pilot descends to the designated spot, aligning with the deck before setting down.

Download on the wing and loss of translational lift significantly increases power required as the aircraft slows through about 20kts. Considerable additional TCL is required to arrest the descent. This is more pronounced in steep approaches (glideslope greater than 3°). Some torque margin must be retained for this purpose.

As with any rotorcraft formation, downwash from the lead V-22 can affect the handling qualities of the wingman, especially within two rotor diameters and in landing. The lead's rearview mirrors are helpful in verifying proper wingman position. The landing order is from the front of the formation to the rear to reduce adverse effects.

Propulsion

The V-22 is equipped with two Rolls-Royce AE-1107C turboshaft engines. These are rated at 6,150shp (4,585kW) at 103.8% Nr at sea level conditions, uninstalled. In the V-22, losses on the order of 10% are experienced in hover through accessories and transmission equipment. It features a 14-stage axial compressor (first 5 stages having variable guide vanes), an annular combustor, a 2-stage high pressure turbine, a 2-stage free power turbine, a torquemeter assembly, and an accessory gearbox. The large number of compressor stages reduces stage loading and tip speeds for better tolerance of dirt and sand ingestion, and inlet distortion. The free turbine, without mechanical connection between the gas generators and power turbines, allows the engine to be turned by the starter without engaging the rotor system. This feature is useful for engine washes in which fresh water is flowed into the rotating engine.

Control of the engine is via a dual-redundant Full-Authority Digital Electronic Control system that optimizes performance without undue attention from the aircrew. Its principal purpose is to covert the power demand signal from the FCS into a fuel flow command. However, it also automatically monitors and limits power turbine torque and speed, and gas generator speed and temperature. Starting, health and limits monitoring, torque matching, and fuel efficiency all benefit from the precision offered by the FADEC. Usually the crew's only interaction with the engine is via the TCL. Separate controls for each engine, including two Engine Condition Levers (ECL), are located in the engine control panel on the overhead console. The ECL selections are CRANK, START, and FLY. Although simply switches, the ECL replicates familiar overhead levers found in helicopters.

The Torque Command Limiting System prevents proprotor mast over-torques, adjusting the torque produced by full TCL based on the mode of flight and conditions. If an engine failure is detected, power is immediately and automatically increased on the remaining engine to the SEO contingency power rating of 6,834shp (5,096kW). This compensates for the reduction in blade pitch to maintain rpm that would otherwise also produce a loss in thrust or lift. The cockpit overhead console INTERIM POWER switch/light provides a mast torque increase up to 109% for 103.8% Nr at full TCL and is commonly used to provide additional lift during heavy weight or hot day takeoffs. It is only available at nacelle angles greater than 80° and below 50kts. Interim power can be armed for automatic activation when these conditions are met.

The V-22 drive train is comprised of two proprotor gearboxes (PRGB), two tilt-axis gearboxes (TAGB), the midwing gearbox (MWGB), and an interconnected drive shaft (ICDS) system. Should the torque of one engine fall below that commanded, uniform proprotor rpm is maintained via the ICDS that runs through wing and connects the engines. Apart from delivering torque to the proprotors, the drive system also distributes engine power to the various systems and components required for flight and mission accomplishment. The transmissions are equipped with an emergency lubrication system that ensures a minimum 30 minutes cruise flight after loss of primary lubrication. They are also pressurized to prevent the intrusion of contaminants.

The PRGB transmit power and provides a 37.8:1 engine-to-proprotor rotational speed reduction. It is located between the engine and the proprotor in the forward portion of the nacelle. The PRGB is connected to the TAGB by a pylon-mounted drive shaft. The PRGB also serves as a forward mounting point for the engine though a gimbal assembly, and provides a load path from the proprotor to pylon support structure.

The left and right TAGB are mounted on the outboard end of the conversion spindle (nacelle rotation point off the wing) in the mid/upper portion of each nacelle. Each consists of a bevel gearbox and an accessory gearbox. The bevel gearbox changes the angle of the drive to transmit power from the PRGB to the TAGB in the opposite nacelle. This connection synchronizes left and right proprotor speed via the ICDS and allows power transfer in either direction during SEO operation. It also performs a 1.81:1 speed reduction between the pylon-mounted drive shaft and the ICDS. The accessory gearbox provides power to engine ancillary components.

The two engines are housed in the two tilting nacelles. The nacelles can be raised manually for maintenance, particularly with the aircraft in the hangar deck of a vessel when the APU cannot be operated. Various openings allow outside air to flow into the nacelle to cool components, assisted by the transmission compartment blower. Lines running into the nacelle from the wing, such as for hydraulics and fuel, are fitted with swivel joints. A fairing between the nacelle and the wing encloses the conversion actuator that performs the tilting.

The nacelles incorporate panels for ready maintenance access to the engine and gearboxes, four hinged panels on the outboard side of the nacelles being primary. The engines are frequently serviced with the nacelles tilted to 0° and the proprotors rotated to ensure that they do not strike the deck. With the nacelles in this 'maintenance position' the lower two outboard access panels can serve as platforms for maintenance personnel, reducing the need for work stands.

Above: **Unpowered flaperons can droop farther than their normal maximum deflection angle. The opening in the mid-wing fairing aft of the flaperon is the SDC exhaust. Lower photograph provides a good view of the slot above the flap.** Author; Jay Miller

Below: **The large and heavy engine nacelles were analogous to mounting a Bell 206 JetRanger helicopter at each wingtip.** Author

Far left: **The junction of the wing and nacelle shows the fairing around the conversion actuator, the wing fence, and intake markings.** Author

Left: **Engine is accessed through a variety of hinged and removable panels.** Jay Miller

Bottom: **The base of the conversion actuator can be seen inside the fairing that covers it regardless of the rotation angle.** Author

The Engine Air Particle Separators (EAPS) in the inlet is important for operations in dusty and sandy conditions. This is a scavenging device at the bottom of the inlet just ahead of the engine face that draws in heavy particles from the airflow and dumps them overboard.

The nacelles also house the Infrared Suppressors. The IRS reduces the infrared signature of the aircraft by mixing the hot engine exhaust with cooler air drawn from within and outside the nacelle. It also reduces the intensity and length of the heat plume extending from the nacelles to lower this hazard during ground operations. The end of the IRS and nacelle include a Coanda deflector system. This blows engine bleed air perpendicular to the exhaust flow, deflecting it away from the fuselage. This operates only with the aircraft on the ground and has the effect of reducing the velocity and temperature of the exhaust flowing towards the fuselage.

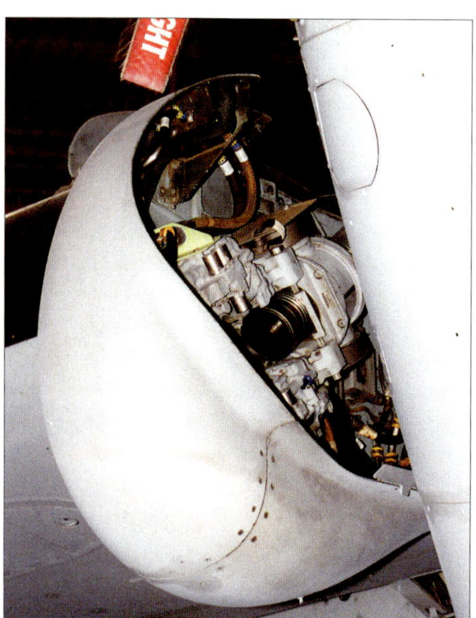

The ICDS runs through the MWGB housed within the wing cavity above the cabin. A number of vital accessories and the APU are mounted to the MWGB. This gearbox, accessed by doors in the overwing fairing, affects a cross-shaft angle change required by wing sweep and dihedral. Drive shafting between the two TAGBs and MWGB consists of ten ICDS segments. Among the MWGB accessories is the rotor brake. This normally stops the proprotors in 10 seconds after engine shutdown and application of the brake. An emergency stop from 100% rpm is possible in 20 seconds. Rotor locks prevent proprotor rotation on the ground in winds up to 45kts and even with one engine idling.

The 300-shp (225kW) APU is a centrifugal flow gas turbine that burns aircraft fuel. Inlet air is drawn from the MWGB compartment and the exhaust is on top of the aircraft. The unit may be started and engaged with the proprotor drive train operating or still. The APU drives the MWGB as motive force for minimum electrical, hydraulic, and pneumatic power for ground checkout, engine starting, BFWS, and environmental control. This eliminates the need for ground support equipment away from the main operating location. The APU control panel is in the cockpit overhead console.

The Vibration, Structural Life, and Engine Diagnostic (VSLED) system supports maintenance analysis of many systems normally requiring detailed breakdown inspections at fixed intervals. Instead, checks and corrections are performed when the recorded data indicates it is required. Monitored are the engines, powerplant structural life, rotor track and balance, structural strain, and vibration diagnostics. VSLED provides data for engine fatigue life determination and vibration trend computations to reveal system degradation.

Fuel System

The principal fuel tanks are two 478 USgal (1,809 lit) cells in the forward portion of each sponson. A 316 USgal (1,197 lit) cell is located in the aft portion of the starboard sponson. Initially intended in FSD as a tank for the CV-22 and available as a kit for the MV-22, it was adopted as common equipment. Fuel is fed to each engine via an 88 USgal (334 lit) feed tank just inboard of each wingtip with crossfeed/crosstransfer fuel lines and valves. The system provides a 10-sec negative-g capability via a small reservoir in each feed tank. Should fuel flow from the sponsons be interrupted, the feed tanks provide approximately 30 minutes flight time. Each has a suction lift pump to draw fuel from the sponson tank on its side of the aircraft, and each engine has an engine-driven suction pump to draw fuel from the feeds. A boost pump in each sponson is normally used above 10,000-ft (3,048-m) altitude to transfer fuel to the feeds, in the event of suction pump failure, or during fueling/defueling and fuel dumping. The sponson and wing crosstransfer capability ensures that fuel from any tank will feed the operating engine during SEO scenarios. The CV-22 has a 294-USgal aux tank inboard of each feed tank in the wing. In a typical mission profile this extends unrefueled endurance by 1.5 hours. In late 2003 the decision was made to delete the aft sponson tank and add two 138 US gal (522 lit) tanks to the wings to improve cg management. This would see MV-22B fuel capacity increase to 1,724 from 1,448 US gal (6,526 from 5,481 lit).

The V-22 features accommodations for up to three internal aux tanks installed on the cargo deck for self-deployment. These allow the aircraft to reach to the farthest point on the globe within two days with only ground refueling. For the CV-22, the cabin aux tanks (CAT) are also used for longer-range operations. The palletized CAT units can be loaded and rigged in about one hour per tank, and the aircraft fully

Top: **The internal elements of the Rolls-Royce Allison AE-1107C Liberty turboshaft engine are shown in this cutaway. The accessory gearbox is at the front under the intake to the compressor section.** Jay Miller Collection

Bottom: **The basic element sof the propulsion system contained within the nacelle are shown.** Bell Helicopter

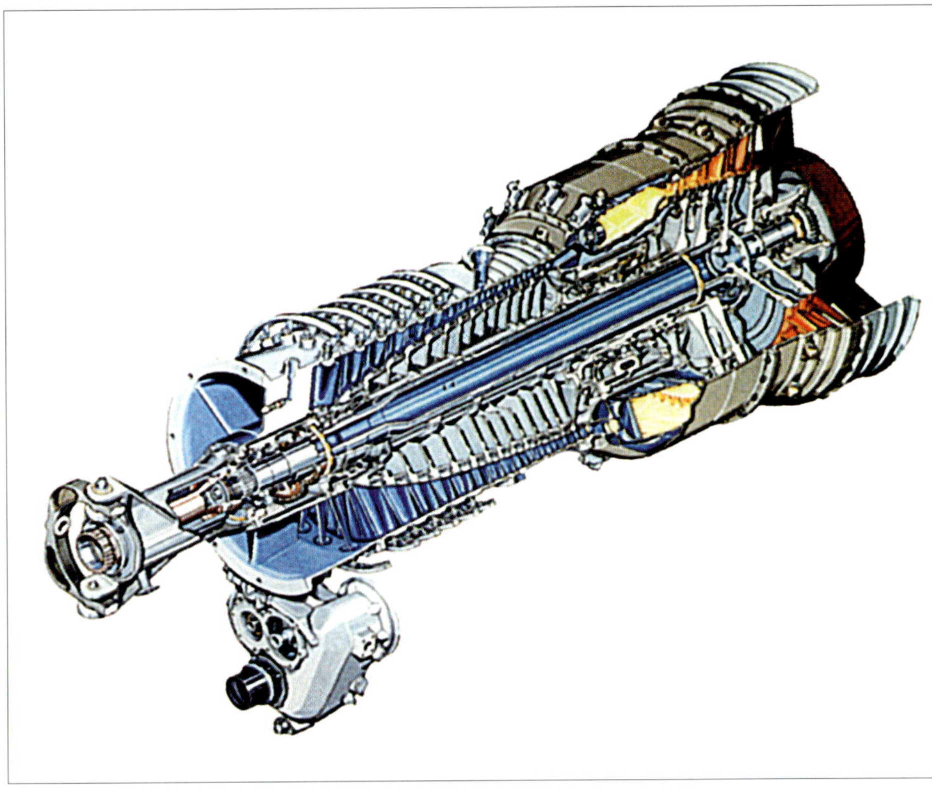

configured for self-deployment in just six hours. Quick-disconnect electrical, fuel and vent lines in the cabin connect to the existing sponson systems. The 800 USgal (3,030 lit) aux tanks (two loadable) were developed but proved large and cumbersome. 200 and 430 USgal (760 and 1,630 lit) units, up to three loadable, appeared more practical and the latter was being bought. The CATs consist of fuel cells suspended in metal frames loaded in tandem on the left side of the cabin, starting from the forward end of the cargo compartment.

All of the fuel (save the CATs) is near the aircraft nominal cg to assist in weight and balance control, and all is outboard of the occupied areas to enhance safety. The fuel cells are made of a synthetic rubberized fabric laced with cord to the aircraft internal structure. They have high tensile strength for crash-resistance against rupture. They have multiple layers of polymer sheeting and self-sealing foam. The foam prevents a dangerous leak if piercing by a 12.7-mm round. The feed tanks are fully self-sealing while the sponson tanks are self-sealing only on their lower third and inboard surfaces. The lower third of the CV-22 aux tank cells are also self-sealing. Self-sealing breakaway valves are fitted at each line-to-tank connection to help prevent spill in the event of a crash landing. The fuel lines also incorporate shutoff values. Refuel lines are scavenged of remaining fuel as a survivability measure in the event of a crash landing or hit by enemy fire. Climb/dive valves (vents) in each tank relieves pressure during ascent and descent to prevent tank rupture, with rupture discs provided for such an eventuality. An On-Board Inert Gas Generating System produces and delivers nitrogen-rich air to fill ullage space and prevent accumulation of an explosive fuel/air mixture for enhanced safety in the event of the tank being pierced by an incendiary round.

The fuel system is controlled and monitored by the Fuel Management Gauging System. This consists of two Fuel Management Units (FMU) that control the fuel burn sequence to ensure the aircraft stays within cg limits. The standard burn sequence is to draw from the aft sponson tank first, the cabin aux tanks (if installed), wing aux tanks (for CV-22), then the forward sponsons, and lastly feed tanks. The computers automatically gauge, feed and transfer fuel to reduce pilot workload, even in the presence of system failures, although some manual backup capabilities remain.

The fuel system can be monitored on the Fuel Status Layer of an MFD, with quantities in each tank and status of the crossfeed/crosstransfer valve shown. The mission computer provides caution at the minimum 'joker' fuel selected in the flight plan and at 'bingo' fuel. All fuel except that in the feed tanks can be dumped through a fuel jettison line from the aft sponson tank, emerging from the bottom of the right hand sponson.

The aircraft is commonly fueled through a single-point ground pressure-refueling adapter in the port sponson and monitored via the adjacent ground refuel/defuel panel (GRDP). Although the FMUs control the system during fuel onload, the GRDP permits some control of the operation and shows system status. Ground refueling can be accomplished with the wing in either the stowed or spread positions. Gravity refueling may also be done through a filler port atop the left forward sponson tank.

Probe-and-drogue aerial refueling may be performed using the probe extending from the nose of the aircraft. The probe has a separate illumination source in the nose of the aircraft for night refueling. Refueling from the KC-135 or KC-10 requires the V-22 to fly at or above 180kts airspeed. The KC-130 tanker has a

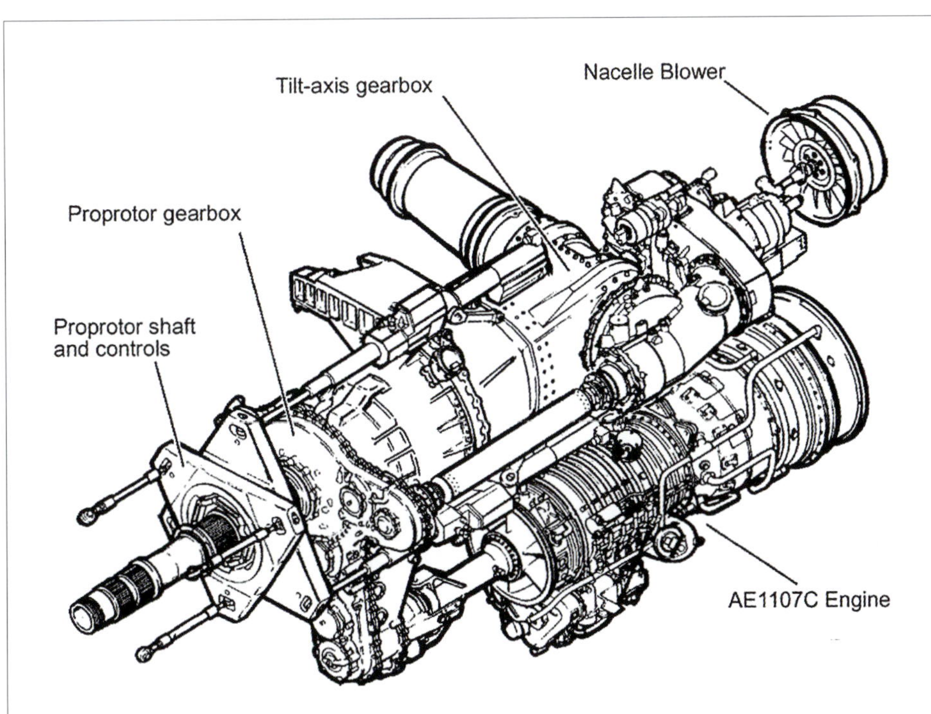

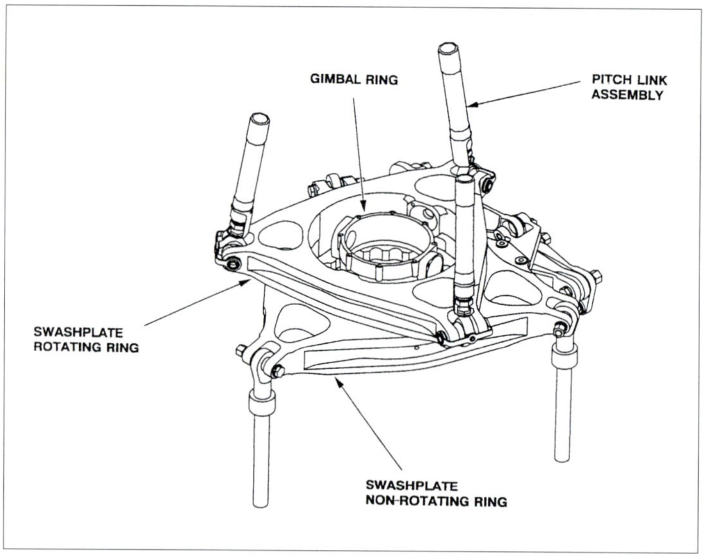

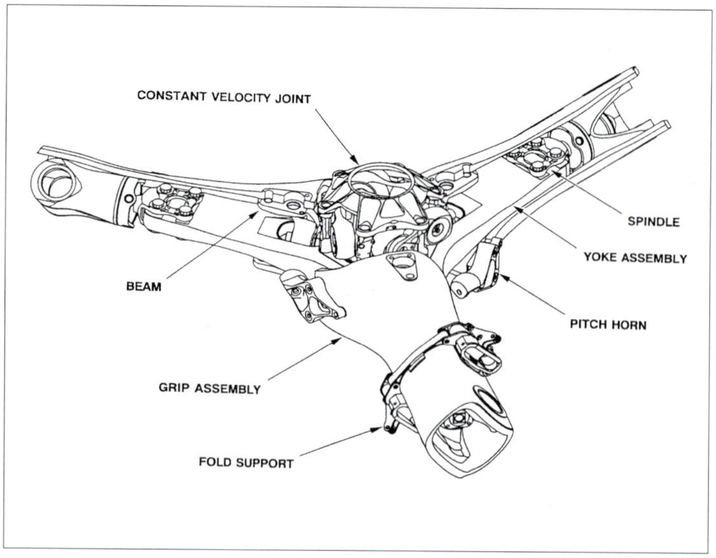

much more compatible AR speed range. A low-speed drogue was under development at time of the V-22 EMD that was expected to improve AR with the KC-130. Initial flight trials demonstrated that the original AR probe was too short for some pilots to comfortably observe probe/basket contact. Late in EMD it was decided to use a lengthened probe on the MV-22Bs that would be supplied as a kit and would probably only be installed when likely to be needed. As originally conceived this kit, installed and checked in about one hour, added 63 lb (29kg). To ensure it would be out of the field of regard of the radar, the CV-22 probe was designed to be fully retractable, with only a small portion exposed at the aircraft nose.

The V-22 was designed to be refueled while hovering off a vessel at sea using the Hover In-Flight Refueling (HIFR) capability. A fuel hose would be drawn aboard through the main cabin door using the rescue hoist and then connected to the HIFR receiver just inside the door. The HIFR feature had not been tested to date of writing and it was possible the capability would be deleted. It was suspected more fuel would be burned hovering than that drawing aboard via the low flow rate hose.

Landing Gear

The tricycle landing gear includes nose and twin main struts fitted with two wheels each. The extend/retract actuators also serve as drag struts while the gear posts serve as vertical shock struts. The nose gear retracts aft and up. The main gear retract forward and up into the sponsons. Each main gear is designed with a two-stage gas/oil shock absorbing system to cope with the impact loads of a 12fps (3.7mps) sink-rate hard landing and 24fps (7.3mps) crash landing, and allows the crossing of a 4in (10cm) bump during ground roll. The tires provide adequate flotation to permit VTOL from exceedingly soft surfaces.

Gear actuation is via hydraulics while a pneumatic emergency blow-down system, one-shot only, is provided in the event of an electrical or hydraulic malfunction. A 2,800-psi (193-bar) nitrogen bottle located in the starboard sponson facilitates blow-down. The pressurized gas unlocks and powers the actuators.

Well doors are mechanically linked to the gear and thus are opened and close via the gear cycle without separate actuators. The opening stroke of the nose doors carries the upper portions into the well to reduce the depth of the exposed portion for easier access to components and towing freedom. The forward main gear doors only open during the gear extend and retract cycle. However, as a maintenance feature the doors can be dropped to the ground, or lowered farther should the aircraft be jacked up, to provide access to the gear well. The aft triangular segment of this door is hinged such that it can be rotated forward against the bottom of the door to lay flat against the ground when the door is manually dropped down. This allows the door to open farther and provides pads on which the door can rest without damage. The larger aft main gear doors can be raised and locked such that it serves as a maintenance platform. For this purpose the door has no-skid material on the outer surface.

Landing gear extension is automatically inhibited above 140kts and warnings activated. Below 60kts airspeed and 200ft radar (60m) altitude, warnings will activate if the gear is not down. Pressing the mute switch on the landing gear panel silences the aural warning, but the gear handle light will continue to flash. Aural warnings are digitized recordings of Barbara (Barb) Smith, once deputy program manager of the V-22. Instead of the usual pilot complaints about 'Bitching Betty', they substitute 'Bitching Barb'.

Top left: **Elements of the swashplate assembly.** Bell Helicopter

Top right: **The proprotor hub assembly is detailed.** Bell Helicopter

Left: **Unlike most helicopters, the entire hub assembly is enclosed in the spinner for reduced drag.** Author

Right: **The V-22 proprotor blades are constructed almost entirely of composites, but still have a number of parts.** Bell Helicopter

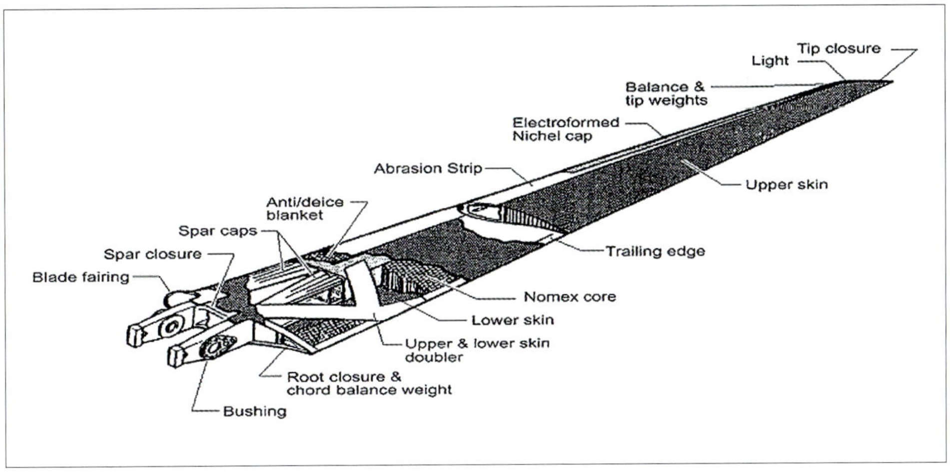

Bottom left and right: **The intakes at the end of the nacelle provide cool air to mix with the engine exhaust in the IR suppressor.** Author (left); Jay Miller (right)

On the ground the hydraulically powered steerable nose wheel can be turned through 70° left or right of centerline as commanded via the rudder pedals. The nose wheels are commanded to the trail position automatically upon gear retraction. With power steering selected off, the nose gear becomes fully castering – an essential feature for close quarters towing on crowded decks. The nose gear can also be locked in the center position for VTOL. In cases of run-on landings, care must be taken not to engage power steering until the nose gear wheels are on the landing surface. A sudden angle change on the wheels while still light on the nose tires can overload the gear once the aircraft has settled and cause a nose gear collapse.

Because of the low clearance under the nose, the nose gear locking pin is inserted by pulling on a handle installed on the starboard side of the nose. The nose gear features a nose hike function that raises the nose, allowing easier access to the gear for maintenance or towing. An external pneumatic source charges the hike system via a valve on the port nose cheek fairing, hyper-extending the shock strut about 15in (38cm).

The main wheels are fitted with carbon brake assemblies consisting of two rotors and three stators. Controlled via pressure on the toe brakes atop the rudder pedals that command hydraulic pressure to the wheel brakes, they provide positive stopping for a refused rolling takeoff from 60kts at 55,000 lb (24,948kg) GW. Pulling the T-handle at the aft end of the center console and depressing either pair of tow brakes sets the parking brake. It is release by depressing the right brake pedal. The aircraft is not equipped with an antiskid system, although a comparable system was under consideration at the end of EMD.

Avionics

A pair of redundant Advanced Mission Computers (AMC or just MC) and three data buses are the heart of the digital avionics system. The software that integrates all the avionics is occasionally altered and reloaded into each aircraft to incorporate improvements or additional capabilities. Avionics components are distributed within the aircraft to the extent possible to reduce system vulnerability.

BIT monitors approximately 850 individual system tests. The Warning, Caution, and Advisory system provide indications on the cockpit MFDs while others are displayed on panels dedicated to specific systems. A Crash Survivable Memory Unit (flight data recorder) is located in the forward end of the wing fairing above the cabin. The unit continually monitors and records selected flight and system parameters.

The AN/AAQ-27 Forward Looking Infrared turret is mounted under the nose centerline. This has an indium antimonide staring array detector operating in the 0.3-0.5 micron energy range, minimizing obscuration from clouds and smoke. The two-axis, gyro-stabilized, pivoting turret gimbal limits are +14° to -23° in elevation, ±30° in azimuth, and it possesses a range of 910-60,000ft (277-18,288m). In its stowed position the optical window is hidden from view. The FLIR, used primarily as a night/adverse weather pilotage aid, provides an image of the surroundings using objects' IR signature. The green-white image is available on the MFDs, with a number of MFD layers providing control of the system and display symbology. Operating modes are Standby, Fixed Forward (turret staring forward), Manual (operation via the track handle), Point and Track (turret rotates automatically for a constant scan of a designated area), and Scan. The track handle has a FOV/Scale Control switch that is used to set the wide, medium or narrow field of view.

The CV-22 has an AN/APQ-186 multi-mode radar in a pod projecting from the port side of the aircraft's nose. The MMR's primary function is to permit low altitude manual TF/TA flight during day, night, and adverse weather to avoid detection by enemy defenses. TF flight at altitudes (termed set clearance plane, SCP) of 100, 150, 200, 300, 500 and 1,000ft (30-305m) are selectable. The unit provides ±40° azimuth and +23/-40° elevation scan. Slant range is 750-60,000ft (229-18,288m) at normal power for TF, 750ft (229m) to 15nm for TA. Radar returns can be viewed in many formats on the MFDs, with a number of MFD layers providing control of the image and overlaid flight guidance symbology.

The radar detects terrain ahead and, using other aircraft inputs, provides pilot cues in flight director symbology to ensure a safe climb over obstacles at the SCP. Many functions and warning features work to ensure safe flight. The TA submode simply provides a display of terrain at and above the aircraft's altitude to provide some terrain masking without high-workload TF flight. The MMR also has a number or submodes for weather detection, ground mapping, reduction of emissions to help avoid

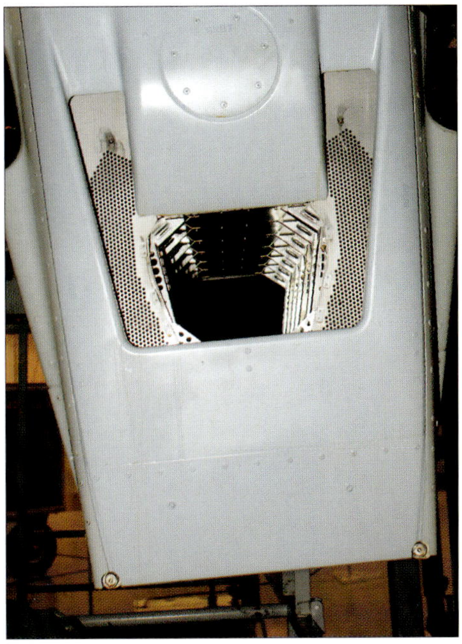

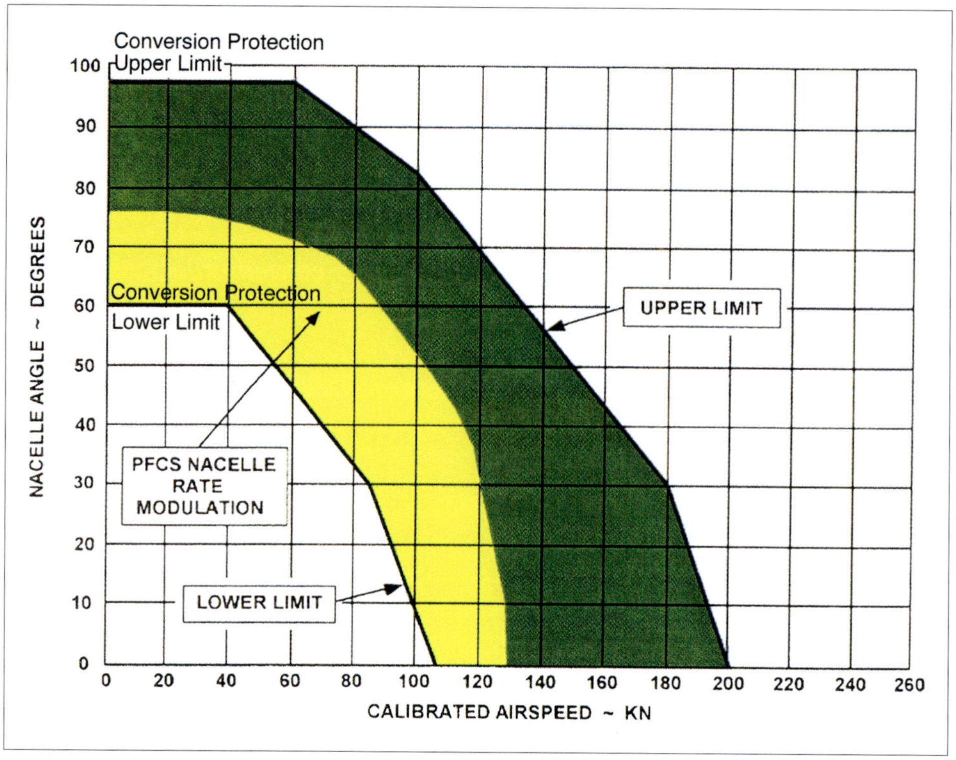

detection, and for interrogation, reception and display of transmitting radar beacon transponders (usually marking an LZ). The radar's Air-to-Ground Ranging mode provides an automatic search, acquisition, range tracking and lock-on of designated terrain features from 9,120-60,000ft (2,780-18,288m) range. These slant range values are helpful for more accurate FLIR inertial navigation system update and store functions to improve nav accuracy.

The mission data loader (MDL) permits pre-planned mission data, prepared off the aircraft using a computer workstation, to be rapidly loaded into the aircraft MCs via a cartridge inserted into a port in the cockpit right side console. This will include radio frequencies, aircraft weight and cg, reference locations (such as threats) or navigation waypoint, and other information that will configure the aircraft's electronic systems and display vital flight planning information during the course of the mission. VSLED data is downloaded to the MDL for post-flight review, although some data is available in-flight.

Communication/Navigation Systems

The MV-22 is equipped with two AN/ARC-210(V) radios possessing voice and data communication with satellite communications (SATCOM) on VHF and UHF channels with anti-jam, and encryption features. These radios also provide FM homing and UHF Automatic Direction Finding. The usual VHF and UHF navigation radio and instrument approach functions are included. The Identify Friend or Foe transponder features embedded encryption, with unique aircraft data for the secure Mode 4 loaded using a fill port on the face of the unit. The frequencies to be used on a mission and assigned to individual radios, called the comm plan, are provided to the MCs via the MDL. The plan is then accessed via the CDU/EICAS keyboard and displayed on the CDU layers where radios and frequencies may be selected and the plan altered as required. The remote frequency indicator selector in the glare shield allows easy changing of frequency. A backup independent control head for one of the radios is provided in the overhead console in the event of system failures. An emergency key on the CDU/EICAS keyboard brings up a special CDU layer with emergency communication frequencies and IFF channels immediately available. It also provides a ZEROIZE key that will erase all classified data from the transceivers and comm plan.

Top left: **A look up the 'tailpipe' of the V-22 shows the Coanda assembly tubes on either side of the exit face.** Author

Above: **The V-22 conversion corridor is shown in simplified form, with small avoid regions at certain flight conditions left off for clarity.** Author

Left: **Departure from the ship is captured as, after rising to a hover, the aircraft translates over the water and accelerates away during conversion. Seen is aircraft 10 during sea trials aboard the USS *Iwo Jima* in January 2003.** NAVAIR

Right: **The basic fuel configurations for the V-22 (prior to Block B changes) is shown with the MV-22 at left, CV-22 in the middle, and CV-22 with notional cabin aux tanks (CAT) at right. CATs would also be used in the MV-22 for self-deployment or special missions.** Bell Helicopter

Right middle: **The fuel dump line is found under the aft portion of the right side sponson. Note the depth of the cargo ramp.** Author

Right lower middle: **The Ground Refuel/Defuel Panel is revealed at the top left of the area uncovered by the raised panel in the left aft sponson of FSD aircraft 1 during a manufacturing test. The panel was subsequently divided in two.** Jay Miller

Bottom left: **The AR probe tip will be standard regardless of the nature of the boom, retractable or fixed.** Author

Bottom right: **Aircraft 7, following the modification as a CV-22 testbed, has a dummy installation of the fully retractable AR probe with light above.** Author

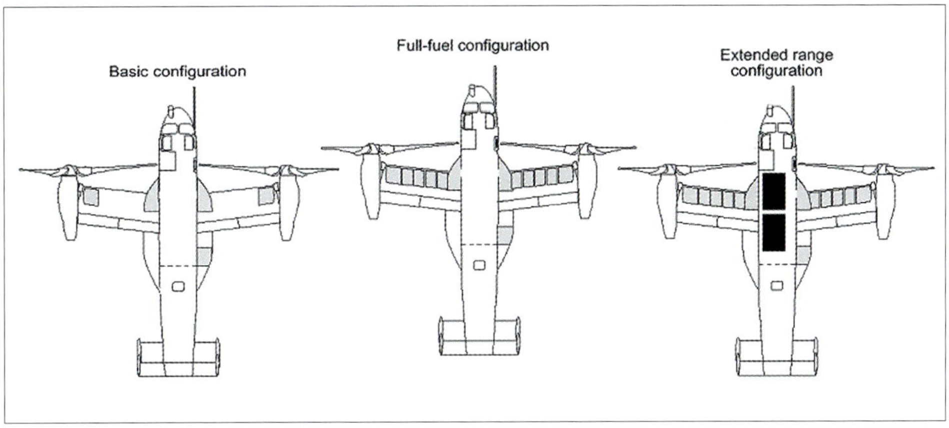

The CV-22 is equipped with two additional AN/ARC-210(V) radios. A troop commander's communication station at the forward starboard side of the cabin has an antenna jack for a man-portable VHF/FM radio and access to the aircraft radios. A data port in the cabin for the troop commander to track mission progress and receive updated information was planned as part of Block 10.

The CV-22 has a Multi-mission Advanced Tactical Terminal (MATT) that receives UHF satellite transmissions of tactical data from various networks with such information as current threat and survivor location. The data is displayed in various formats, including superimposed on the digital map. Downed aircrew survivor information includes identifier and location. Crypto communications data is loaded into the MATT via a fill port at the avionics bay.

The inter-communication system includes six ICS panels with three in the cockpit, one in the left forward cabin and right aft cabin, and one external for ground personnel. These allow communication within the aircraft via headphones with boom microphones and push-to-talk buttons. They also permit monitoring of radio traffic and some transmission capability from within the cabin.

The Crash Position Indicator is an emergency locator transmitter. The unit transmits a homing signal on international distress frequencies when subjected to high decelerations associated with a crash. The rugged, sealed unit, located in the ceiling of the aft cabin, should survive most crash situations. Batteries permit continuous operation in excess of 48 hours.

The heart of the Osprey's navigation system is the triple redundant Light Weight Inertial

Navigation System (LWINS). This is aided by the Miniature Airborne Global Positioning System Receiver (MAGR) to initiate and sustain accurate navigational functions. The Global Positioning System is a constellation of satellites that help the aircraft fix its position on or above the earth with great accuracy. Data from LWINS is processed for position and navigation guidance to the crew and autopilot.

The Digital Map System provides current aircraft position and navigation data overlaid on one of several dynamic color moving map display presentations. This draws upon a vast map and terrain library in the unit's memory. Navigation and threat symbology options are available. The flight plan can be edited on the DMS by moving waypoints using the track handle and displayed cursor. The system can provide attributes of displayed threats as well as compute and display the best route between and around them. The database can also be provided with up to 100 scanned images, such as attack plans and reconnaissance photos, for display to aid mission execution.

Electronic Warfare Systems

The MV-22B is equipped with an AN/APR-39A(V) Radar Signal Detection Set, AN/AVR-2A Laser Detector Set (LDS), and AN/AAR-47 Missile Warning System (MWS). Threats are detected, identified, and the crew alerted through aural and visual means. The threats are shown in the associated quadrant around the aircraft on the Radar Signal Indicator (RSI) mounted in the instrument panel. The MWS senses infrared rocket engine plumes from missiles. Four MWS optical sensors are installed in the four quadrants of the jet; one in each of the fairings on either side of the nose and one in each fairing at the aft end of each side sponson. Like the MWS, the LDS has sensors in the four quadrants to detect laser energy aimed at the aircraft as part of an enemy targeting system. The AN/ALE-47 countermeasures dispensing system (CMDS) has two flare/chaff dispensers. One dispenser is installed

Top: **This photo montage shows a goodly portion of the V-22 landing gear retract cycle sequence. Extension is the reverse of that seen here.** Paul Shank

Left: **The main landing gear retract forward.** Jay Miller

Photographs on the facing page:

Left: **The bulged main gear forward door segments are closed except during door cycle and ground maintenance. The large strake above door creates favorable flow over aft portion of sponson while also doubling as the VOR/ILS localizer antenna.** Author

Right: **The no-skid strips on the main gear door (looking forward) are important when raised to serve as a work platform.** Author

in the aft portion of each sponson and fires multiple types of rounds. A cockpit control unit allows selection of CMDS functions as well as displaying the number and type of rounds remaining. Expendables from the CMDS are fired using one of six pre-programmed firing sequences. Manual firing uses a switch on the TCL or a dispense switch on a 12-ft (3.7-m) chord plugged into to one of three connectors in the cabin.

The CV-22 has an additional expendables 'bucket' in each of the aft sponson stations as well as one under the forward fuselage just aft of nose gear door. A second nose dispenser is to be added in Block 10. The CV-22 uses Army and Air Force expendable rounds that have a rectangular or square cross-section vice the Navy's circular cross-section.

The CV-22 is equipped with the AN/ALQ-211 Suite of Radio Frequency Countermeasures. It automatically geo-locates and characterizes detected threats for presentation on the moving map display, and determines the best means of meeting the threat. Electronic emissions are sensed via four receive antennas, one on either side of the lower forward fuselage and one on either side of the aft fuselage. SIRFC also considers threat information provided by the MWS and MATT, and those loaded by the aircrew into the DMS. The SIRFC may choose to jam the threat at the appropriate moment using its four transmit antennas; one on either side of the forward fuselage and one at the end of each vertical stabilizer. It may choose to command the appropriate decoy flare or chaff expendables to be fired from the CMDS. Threats are also presented on the Dedicated Electronic Warfare Display (replacing the MV-22 RSI) with bearings to SIRFC-processed or MWS-detected threats, and via aural warnings. The crew may manually command SIRFC via MFD layers.

Hydraulics System

The Osprey features three independent 5,000-psi (345-bar) hydraulic systems. Pumps on the two engine TAGBs provides pressure to Systems No 1 and 2 while the No 3 pump is mounted to the MWGB. A flight control module for each pump, directed by FCCs, controls the system and ports pressurized fluid to the various actuators. Four thermal control modules are used to rapidly warm the fluid during cold weather operations. The system components and lines are separated to help ensure survivability in the event of damage. A FCC can detect a leak within 0.3 seconds and take action, as determined by switching logic, to isolate the defective area and prevent loss of vital fluid. Appropriate valves are closed and fluid rerouted to maintain essential functions.

Systems No 1 and 2 are the primary power sources for flight controls while No 3 provides power to some control surfaces and to utility systems (landing gear, steering, brakes, ramp and door, wing stow, and so on). System 3 is the source of hydraulic power during ground operations, powered by the APU or ground hydraulic cart. It will pressurize all systems until engine start. The failure of No 1 or 2 in flight will result in 3 taking up the load in a dual-redundant fashion. Vital control of the swashplate and conversion actuators is triple redundant with No 1 and 2 powering each dual-sided actuator with 3 as a backup for either system. Failure of one system should not prevent mission completion, and loss of two systems would not preclude safe flight. Failure of No 3 would render the utility systems (such as nose wheel steering) inoperative and require emergency procedures. The hydraulic system is monitored on the appropriate MFD page.

Electrical System

The V-22 electrical system consists of constant frequency generators (CFG) and variable frequency generators (VFG) for 115/200 VAC power, solid-state regulated converters for DC power, four AC and six DC electrical power buses, and a battery.

The two 40 kVA variable input speed CFGs supply 400Hz power to AC buses 1 and 2 as the primary power to the aircraft. The No 1 CFG is located on the MWGB and is driven by the drive train turned by the engines, or the APU for ground operation. The No 2 CFG is located on the right TAGB. The two 50/80 kVA VFGs are secondary AC sources and principally supply power to environmental buses. The No 3 VFG is located on the left TAGB while the No 4 VFG is located on the MWGB. External AC power is supplied during ground operations via an external power receptacle located forward of the left sponson. Four AC 'house power' receptacles are distributed throughout the cabin, above the troop seats, to power cabin aux tanks, medevac equipment, and carry-on gear.

Three transformer-rectifiers or converters supplying 28 VDC power. Converters 1 and 2 are 200 A units that are powered from AC buses. One 24-amp/hrs sealed and heated acid battery is located in the cabin ceiling and is accessible from within the aircraft. It is primarily used in starting the APU and powering the maintenance and unswitched buses when AC power is unavailable, but also supplies 20 minutes of emergency DC power in flight.

In the event of failure of one or more power sources, the system is designed to automatically reconfigure by transferring buses from failed sources to the remaining sources. Most electrical system status and controls are accessed via the MFDs. Individual circuits can be disconnected by pulling the appropriate circuit breaker on panels throughout the cockpit and cabin.

Environmental Systems

The V-22 has been designed to performs all its normal operations in the wide spectrum of environments that may be encountered anywhere on Earth, either at sea or ashore. This includes salt spray what can accelerate corrosion, icing, lethal chemical or nuclear conditions, extremes of temperature and humidity, and blowing sand.

The environmental control system supplies conditioned air for crew comfort and to onboard equipment. This allows the aircraft to

Left: **Nose landing gear. FLIR turret is rotated to the stowed position.** Jay Miller

be operated at ambient temperatures of -65 to 125°F (-54 to 52°C). The system is designed to maintain the cockpit at 60-80°F (16-27°C) and the cabin at 40°F (4°C) to 10°F (-12°C) above outside air temperature (OAT). A cold weather kit consists of cabin sidewall and overhead thermal blankets, and a curtain across the ramp area. It can be installed when the aircraft is to conduct extended operations in cold climates.

A shaft driven compressor (SDC) on the MWGB supplies the air flow for the ECS. The intake for the SDC is on the port wing-to-fuselage fairing. Alternatively, compressed air can be pumped aboard from a ground pneumatic power cart via a receptacle on the port sponson. The ECS controller governs operation of the pneumatic portion of the ECS.

An Environmental Control Unit (ECU) conditions the compressed air from the SDC. The unit is housed in the aft portion of the port sponson with a flush ram air intake atop the sponson to assist air cooling. The ECU removes excess humidity, large particles, and conditions the air with an air cycle machine to the requested temperature. The ECU air is then distributed to the cockpit, side and chin window defog outlets, and cabin via ducting and according to the selected levels. Priority is given to conditioning the cockpit first, with whatever remaining capacity directed to the cabin. A series of gaspers in these spaces attempt to direct the air optimally. An emergency vent fan can be used to draw unfiltered and unconditioned air from the port sponson and introduce it to the ECS ducts.

Operation of the ECU is supposed to provide a positive pressure differential between the inside of the aircraft and the outside atmosphere to prevent the intrusion of contaminants. The aircraft is not otherwise pressurized. Should the interior pressure exceed a maximum differential with the outside, air is automatically dumped overboard via spring-loaded poppet valves. The cockpit overflow valve can

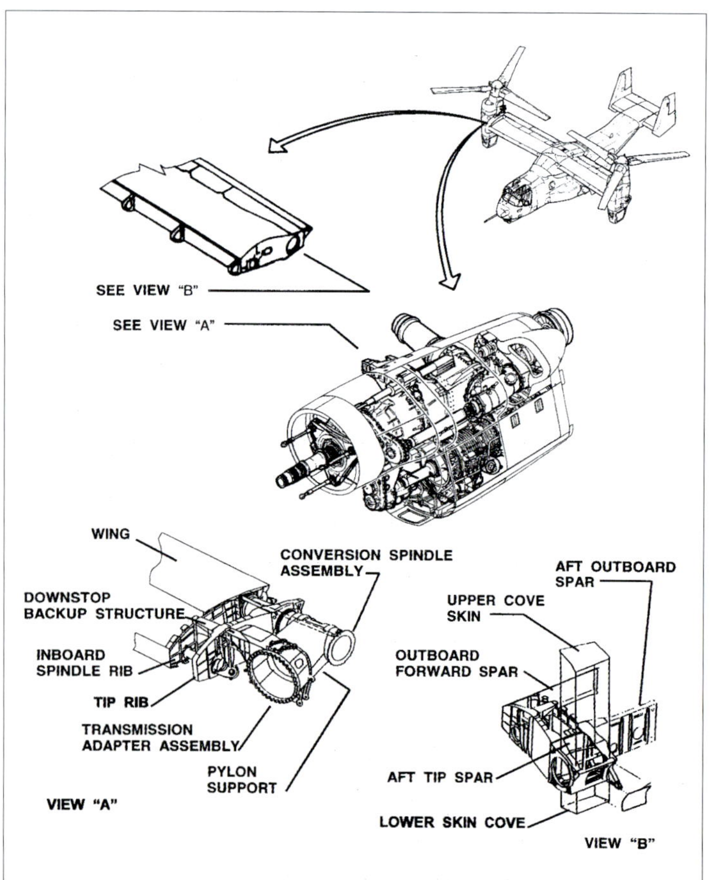

Below left: **The manner of mounting and rotating the heavy engine nacelles and proprotors from the cantilever wing involves stout wingtip structure with mating pylon support and transmission adapter. Note that the engine is just one small element within the nacelle.** Author's collection

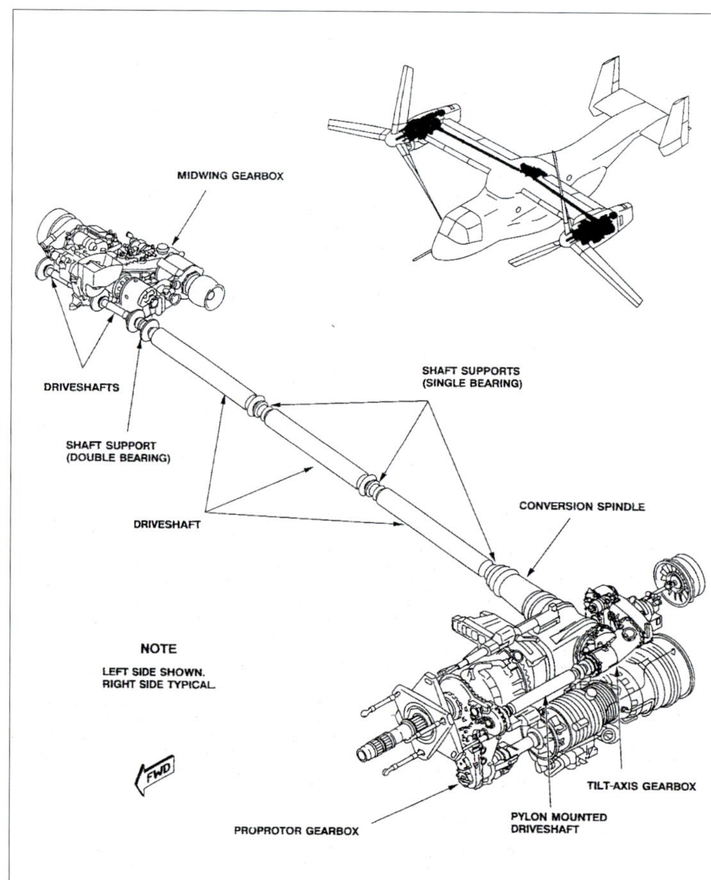

Below right: **The drive shaft connecting the two proprotors passes through the spindle and transmission adapter to the TAGB that turns the rotation through 90°. The short pylon-mounted drive shaft connects the TAGB and the PRGB.** Bell Helicopter

be operated manually for such emergency situations as smoke and fumes evacuation. A NBC filter kit can be installed on the ECU to provide cleaned air during 'buttoned-up' operations in a contaminated area. However, during EMD it provided exceedingly difficult to prevent leaks and sustain necessary overpressure for the capability to function properly, and the NBC kit was considered for deletion. Three NBC aircrew protective suit vent receptacles are also found in the cockpit to provide pressurized air to the suits when worn, and four more receptacles can be installed in the cabin.

The avionics cabinets on either side of the forward cabin draw in ambient air through two particle separators. The first stage of these separators is a pair of flush inlets visible outside the aircraft. The second stage is a filter material to remove smaller particles. Fans or ram pressure then circulate the cooling air. The avionics cooling air is also directed to the nose bay equipment (CV-22) and then discharged overboard. Recirculated interior air, or air drawn in from the outside, cools other avionics components installed in the cabin overhead or underfloor. Over-temperature conditions are annunciated to the crew.

The ECU supplies conditioned air to the On-Board Oxygen Generating System (OBOGS). This system creates and delivers oxygen-enriched air (O_2N_2) for crew breathing, reducing the need to carry and service oxygen bottles. Seven OBOGS outlets are provided: three in the cockpit and two installed on either side of the cabin. Depending upon aircraft altitude, the O_2N_2 production is sufficient for four to nine persons. An oxygen bottle is installed in the cabin overhead to provide approximately five or more minutes of oxygen for an emergency descent should the OBOGS fail at altitude. In the case of High Altitude – Low Opening parachute jumps above 12,500ft (3,810m), a supplemental oxygen supply is brought aboard for jumpers' pre-breathing and to top-off their individual kit tanks before jumping.

The Ice Protection System allows the aircraft to be operated in moderate icing conditions. When the temperature is sensed by the OAT probe to be at or below 40°F (4°C), the IPS is automatically activated until temperatures rise above 45°F (7°C). The leading edge of the wing is fitted with inflatable pneumatic de-icing boots for shedding accreted ice. The CV-22's radome was also to be fitted with a deicing mechanism sometime after EMD. Electro-thermal de-icing heating elements protect the proprotor blade leading edges, the rotor root fairings, and the spinners. Engine bleed air provides de-icing for the intake front frame struts and EAPS. The other portions of the propulsion system plus pitot-static and angle-of-attack probes use electrothermal anti-ice. The engine has demonstrated the ability to pass ice shed from the intake and spinner without difficulty. The pilot windshields are electrothermally anti-iced by a conductive film-heating layer embedded within the transparency. The FLIR base and window is also equipped with heating elements. Activation of anti-ice/de-ice systems can produce a 5-10% reduction in engine torque available.

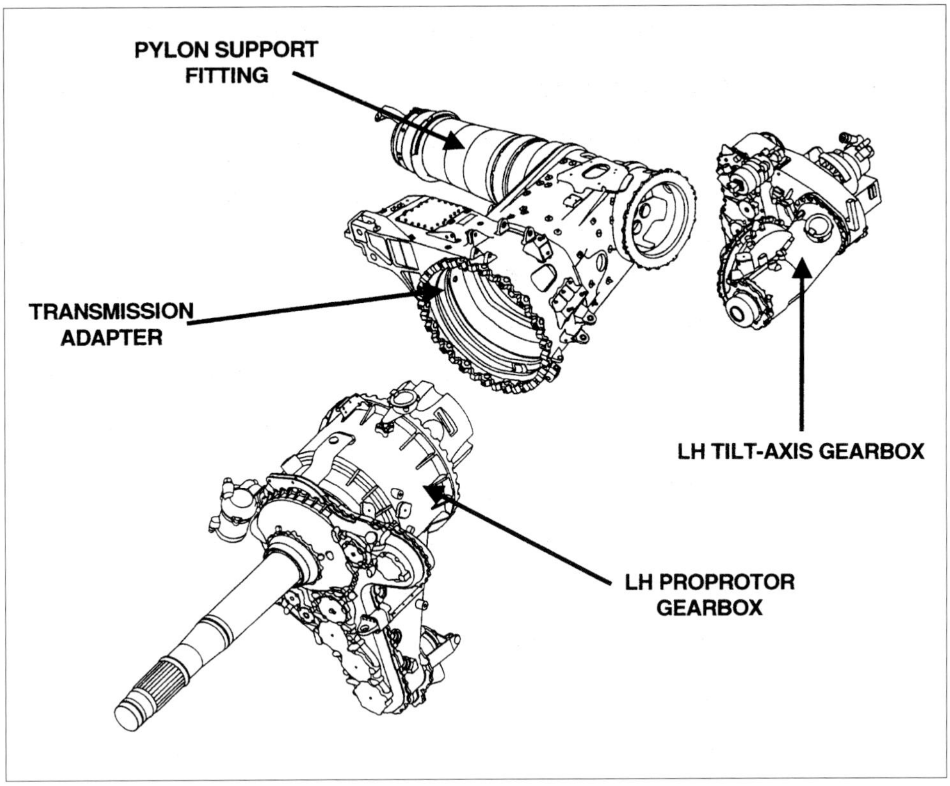

Above: **From the transmission adapter forward is the proprotor gearbox and outboard the tilt-axis gearbox.** Author's collection

Two-speed windshield wipers permit rain removal from the pilot and copilot windscreens and incorporate integral washer spray bars.

The aircraft systems are shielded for protected from the adverse effects of lightning, static electricity discharge, and the electromagnetic environments present aboard ship or likely to be encountered on the battlefield. The composite fuselage structure incorporates a copper mesh laminated into the outer skin. This provides the avionics protection against lightning strikes and the interference of external electromagnetic sources, but also reduces the effects on aircraft systems of an electromagnetic pulse from a nuclear blast. Critical areas around the avionics bay are completely shielded with aluminum skins. The wing and proprotors are also treated to conduct lightning away from sensitive areas and prevent damage to internal structure. The proprotors are capable of withstanding lightning strikes on any part of the blade without catastrophic damage. Static discharge wicks are installed at the trailing edge of many of the control surfaces.

Fire Detection and Suppression

The fire suppression panel in the overhead center console has fire annunciators and an NORM/ARM T-handle for each engine. Fire is sensed by a gaseous pneumatic fire detector loop in each nacelle. A fire bottle in each nacelle is discharged with the T-handle. Electrically operated fire doors on the forward bottom and inboard sides of each nacelle also close.

The aft wing coves, wing dry bays surrounding the fuel cells in each side of the wing, and the midwing area have automatic fire detection and suppression. Fire detection annunciators are provided on the overhead panel.

Lighting

Cockpit lighting is controlled via assigned rheostats. White or NVG-compatible green lighting is selectable. A handheld lamp is installed above each pilot, attached to the overhead panel. Cabin lighting is controlled from a forward panel with a dome light rheostat provided along with a switch to change this lighting from day to blue-green night lighting, and 'night bright'.

Normal position and anti-collision lights are installed for day and night peacetime operations. An upper IR strobe light allows for rendezvous at night while flying on NVGs. Blue/green electroluminescent lights on the fuselage, nacelles and proprotor tips are used for night formation flight. Those on the proprotor blades define the extremities of the vehicle and positioning for landing, while the others support formation alignment or reveal attitude changes.

Two retractable landing/searchlights under the forward fuselage can be extended through 110° from stowed and rotated 360°, as commanded via switches on the TCLs. They have selectable white or IR modes.

V-22 Specifications and Performance

Dimensions

Airframe

Width		
fuselage exterior mold line	7.75ft	(2.36m)
between center of vertical tails	17.75ft	(5.41m)
between outboard moldline of vertical tails	18.42ft	(5.61m)
outboard mold line of engine nacelles	50.92ft	(15.52m)
VSTOL	84.46ft	(25.74m)
APLN	83.92ft	(25.58m)
wing stowed	18.42ft	(5.61m)
Length		
nose to tail	57.33ft	(17.47m)
wing stowed	63.00ft	(19.20m)
Height		
tip of spinner, VSTOL	22.08ft	(6.73m)
rotor tip path, VSTOL	20.83ft	(6.35m)
top of tail beacon	17.92ft	(5.46m)
nacelle rotation axis	12.33ft	(3.76m)
wing stowed	18.25ft	(5.56m)
elevator centerline	9.67ft	(2.95m)
bottom of vertical tail	6.17ft	(1.88m)
bottom of engine nacelle, VSTOL	4.33ft	(1.32m)
cabin floor	2.80ft	(0.85m)
fuselage bottom - nominal	1.50ft	(0.46m)

Wing

Area		
incl flaperons and across fuselage	382.00ft²	(35.49m²)
exposed, including flaperons	301.40ft²	(28.00m²)
Span (engine axis to engine axis, spinner tip)		
VSTOL	46.83ft	(14.20m)
APLN	45.83ft	(13.97m)
Sweep angle	6.00° forward	
Dihedral angle	3.50°	
Chord length	8.33ft	(2.54m)
Airfoil max thickness	1.92ft	(0.58m)
Thickness to chord ratio	23%	

Cabin Interior

Width		
loadable interior	5.67ft	(1.73m)
max	5.92ft	(1.80m)
Length		
loadable interior	20.83ft	(6.35m)
to tail	32.71ft	(9.97m)
Height		
at sidewalls	4.85ft	(1.48m)
at center	5.52ft	(1.68m)
Volume, loadable	738ft³	(20.9m³)
Ramp lowered angle	18.5°	

Proprotor

Blades per hub	3	
Blade length (from center of hub)	19.00ft	(5.80m)
Mean chord length	2.30ft	(0.70m)
Root chord length	2.96ft	(0.90m)
Tip chord length	1.83ft	(0.56m)
Blade twist	47.5°	
Blade area, total (straight taper)	261.52ft²	(24.30m²)
Diameter	38.08ft	(11.61m)
Disk area, total	2,268ft²	(211m²)
Blade		
pitching & flapping angle, max	10.50°	
clearance over cabin, min, VSTOL	4.67ft	(1.42m)
fuselage clearance, min, APLN to cabin	1.00ft	(0.31m)
to sponson	0.75ft	(0.23m)
wing clearance, min, APLN	0.92ft	(0.28m)
Tip distance below fuselage, max APLN	9.50ft	(2.90m)
Nacelle tilt limits (horiz to aft of vertical)	0 to 97.5°	

Control Surface Rotation Limits

Elevator, up	30°	
down	20°	
Rudder, either side of center	20°	
Flaperon, for roll	25° up	40° down
as flaps	72.5° down	

Landing Gear

Wheel track		
outside of outer wheels	15.42ft	(4.70m)
strut to strut	13.00ft	(3.96m)
Wheelbase	25.00ft	(7.62m)
Nose gear steering angle	±75°	
Tip-back angle on ground, max	13.00°	
Roll angle on ground, max	14.40°	

Weights

Empty (objective)	33,140 lb	(15,030kg)
Design gross (60% fuel)	39,500 lb	(17,917kg)
Takeoff gross (VTOL, sea level), max	52,600 lb	(23,860kg)
STO gross, max	57,000 lb	(25,855kg)
Self-deployment GW (smooth field STO)	57,000-60,500 lb	(25,855-27,440kg)

Fuel Capacity (pre-Block B changes)

Total usable (MV), JP-5	1,448 USgal (5,481 lit)	9,850 lb (4,470kg)
Total usable (CV), JP-5	2,036 USgal (7,707 lit)	13,845 lb 6,280kg)
Cabin aux. tank (two loadable), each	430 USgal (1,630 lit)	
Refuel rate, max		
pressure refuel & AR	≈400 USgal/min	(1,515 lit/min)
gravity refuel	≈100 gal/min	(380 lit/min)
Defueling rate, suction or gravity drain	≈80 USgal/min	(303 lit/min)
Fuel dump rate	800 lb/min	(365kg/min)

Power System

Engine

Model	Rolls-Royce Allison AE-1107C turboshaft	
Number	2	
Weight (dry)	971 lb (447kg)	
Power-to-weight ratio	7:1	
Continuous rating, max	5,256shp	(3,919kW)
Rating, max/intermediate (takeoff)	6,150shp	(4,590kW)
SEO rating, contingency power	6,834shp	(5,096kW)
Torque limit	2,153ft-lb	(298kg-m)
SFC, max continuous power	0.428 lb/hp-hr	

Transmission (PRGB output to proprotor)

Transient, VTOL	6,048shp	(4,510kW), 397rpm,
	960,517 in-lb	(132,796kg-m) torque
APLN	5,071shp	(3,781kW), 333rpm,
	960,517 in-lb	(132,796kg-m) torque
Takeoff, VTOL	4,983shp	(3,716kW), 412rpm,
	762,077 in-lb	(105,361kg-m) torque
Continuous, VTOL	4,570shp	(3,408kW), 397rpm,
	725,787 in-lb	(100,344kg-m) torque
APLN	3,831shp	(2,857kW), 333rpm,
	725,787 in-lb	(100,344kg-m) torque

Rotor

Speed	333rpm at 84% Nr	(tip speed 662fps, 202mps)
	397rpm at 100% Nr	(tip speed 790fps, 241mps)
	412rpm at 104% Nr	(tip speed 822fps, 251mps)

Performance

Airspeed		
Design limits	0 to 340kts	
max operating at 15,000ft, 45°F	305kts	
max dive, APLN	280kts	
max cargo ramp open	260kts	
Cruise	250-300kts	
At 3,000ft, 91.5°F	275kts	
max cargo hook doors open	250kts	
CONV (<75° nacelle angle)	220kts	
max cargo hook doors opening	150kts	
max gear extend/retract	140kts	
VSTOL (75-95° nacelle angle)	110kts	
max touchdown	100kts	
max takeoff	90kts	
Sideward translation	40kts	
Rearward translation	20kts	
manual flaps, 20°	240kts	
40°	180kts	
full, 73°	60kts	
Range (sea level, standard day, unrefueled)		
12,000 lb payload, VTO (radius)	140nm	
STO (radius)	313nm	
8,000 lb payload, STO (radius) CV-22	540nm	
5,000 lb external load mission, VTO (radius)	300nm	
Self-deployment, no payload, STO	1,940nm (2 CATs)	
Payload		
Personnel	seating for 4 crewmembers (normal complement 3; pilot, copilot and crew chief) and 24 fully equipped troops	
Medevac	12 litters with 4 medical attendants	
Internal cargo	20,000 lb	(9,072kg)
Airdrop loads	one 3,000 lb	(1,361kg) or
	four 500 lb	(227kg) bundles
Cabin floor & ramp loading limit	300psf	(0.14 bar)
Cargo ramp load, max		
lowered to the ground	5,000 lb	(2,268kg)
horizontal	3,000 lb	(1,361kg)
External cargo, single hook	10,000 lb	(4,536kg)
dual hook	15,000 lb	(6,804kg)
Rescue hoist, max	600 lb	(272.2kg)
Miscellaneous		
Ceiling, max operating, APLN	25,000ft	(7,620m)
CONV	17,000ft	(5,180m)
Load factor, max, APLN	-1 and 3.5G	
VSTOL/CONV	-0.5 to 2.0G	
Bank angle, max, descending	75°	
climbing/level	60°	
Landing sink rate max	12fps (3.66mps)	GW≤46,000 lb
	10fps (3.05mps)	GW>46,000 lb
Slope landing	9° all directions	
Disk loading	17.5-23.2psf	(85.4-113.3kg/m²)
Blade loading, max	178psf	(0.085 bar)
Basic equivalent drag, APLN	32.00ft²	(2.97m²)
Example (MV-22B, 46,000 lb (20,865kg) GW, sea level, standard day conditions)		
Disk loading	20.28psf	(99.02kg/m²)
Takeoff/hover OGE altitude, max		
interim power	4,788ft	(1,459m)
STO distance, min (no wind)		
ground roll	150ft	(46m)
over 50-ft obstacle	280ft	(85m)
Rate of climb, max, APLN	3,165fpm	(16mps)
CONV	2,900fpm	(15mps)
VSTOL	1,720fpm	(9mps)
APLN speed range, wing stall (sea level) to rotor limit (16,800ft)	111kts to 293kts	
Service ceiling	23,585ft	(7,189m)
Normal load factor limits, APLN	-0.86 to 3.01G	
VSTOL/CONV	-0.43 to 1.72G	
Specific range	0.074nm/lb of fuel	
Best range	998nm @ 215kts	
Best endurance	5.2 hours @ 168kts	
STOL landing distance (dry surface, no wind)		
ground roll	1,000ft	(305m)
over 50-ft obstacle	1,600ft	(488m)

Below, left: **The aft fairing on the port sponson of the MV-22B has the radar warning receiver antenna at top and Missile Warning System sensor below.** Jay Miller

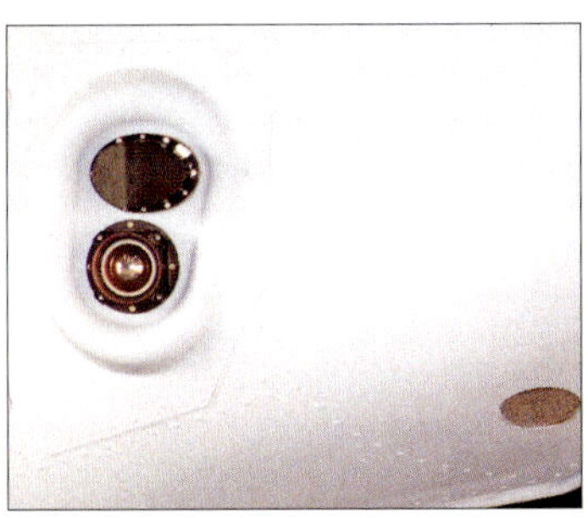

Below, right: **The inlet aft of the door is for avionics cooling air. This is a particle separator made up of a series of holes that induce swirl and separate large particles.** Author

Bottom left: **The MV-22 and CV-22 antennas and other prominent sensors are detailed here.** Author

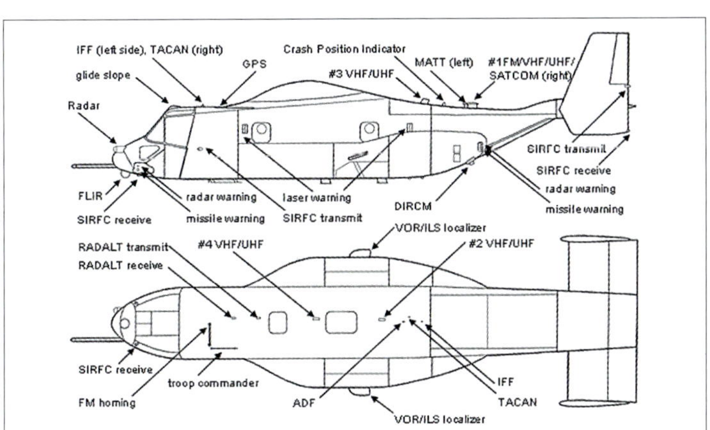

Below: **The V-22 exterior lights are detailed in this diagram.** Author's collection

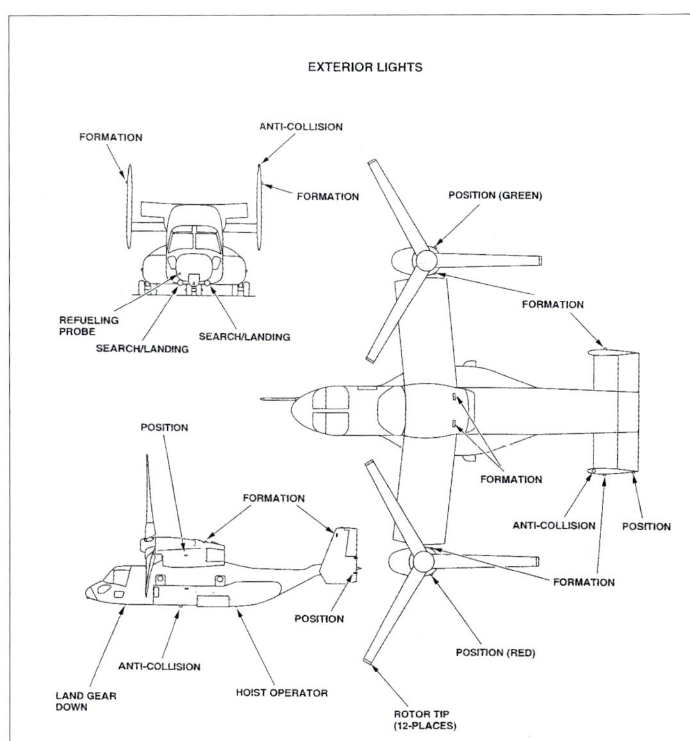

Tiltrotor Gallery

Two views of the Bell XV-3 as displayed at the U.S. Army Aviation Museum at Ft. Rucker, Alabama during August of 1987. The aircraft is configured and painted as it was during its final year of flight test under the auspices of the National Advisory Committee for Aeronautics (NACA), the predecessor to today's National Aeronautics and Space Administration (NASA). Both Jay Miller

Above and inset: **The first of the two Bell XV-15s completed, '702', was shipped to the NASA's Ames, California flight test facility at Moffett Field during November of 1978. The aircraft is seen at Carswell AFB in Ft. Worth, Texas, immediately prior to being loaded into a Lockheed C-5A for the delivery flight to California.** Below: **Cockpit of the second XV-15, '703', during January of 1979 when the aircraft was well into its initial flight test program.** Both Jay Miller

Above: **Bearing NASA markings, the first Bell XV-15, '702', was displayed during an open house at the NASA's Ames, California facility during July of 1985.** Craig Kaston via Jay Miller

Left and below: **The second Bell XV-15, registered N703NA for demonstration purposes, is seen in civil markings painted to show what the aircraft might look like in a corporate configuration. The passenger 'windows' on the fuselage side are actually fake paste-ons. The demonstration took place at Bell's transmission plant in Grand Prairie, Texas, during April of 1995.** Both Jay Miller

Facing page, top: **The first V-22, BuNo 163911 ('01') at the very beginning of its first flight on March 19, 1989. The aircraft taxied out to the main runway from the Bell flight test hangar at Arlington Municipal Airport before beginning its initial hover.** Jay Miller

Facing page, bottom: **View into the cargo hold of the first V-22. Emergency exits provide the only outside view.** Jay Miller

Bell Boeing
01

Bell Boeing

Above and right: **For the rollout ceremony at Bell's Arlington Municipal Airport facility, aircraft 1 was given a two-tone camouflage (FS35237 Medium Gray and FS34095 Medium Field Green) with black markings. The 'ARMY' on the fuselage was misleading as the service had withdrawn from the program earlier.** Both Jay Miller Collection

Below: **Aircraft 12 and 13 set down gingerly in a wet field during operational evaluations of the V-22 during 2000.** Ron Culp

Photographs on the preceding page:

Top: **The initial public demonstration of the first V-22 took place at Arlington Municipal Airport, Texas during March of 1989. The crew consisted of Dorman Canon and Dick Balzer.** Jay Miller

Bottom: **A low pass will surprise observers by how quiet the V-22 is in airplane mode.** Author's collection

COAST GUARD

5435
MARINES

Above: **A large portion of the MV-22 test fleet in 2003 is seen on the flight test ramp at NAS Patuxent River, Maryland.** Navy

Right: **A pair of MV-22B Ospreys cruise in high-speed airplane mode above the ocean.** Ron Culp

Photographs on the preceding page:

Top: **During 'Tiltrotor Day' on September 8, 1999, the first production MV-22B, aircraft 11, and the sole XV-15 operated from the Pentagon's 'River Entrance' parade ground to help bolster understanding and support for the Osprey. Apart from some scorched grass, the event went off without a hitch.** NAVAIR

Bottom: **A team prepares to attach a water tank as a sling load under MV-22B aircraft 13. Noteworthy are the open external load hook access doors on the bottom of the aircraft.** Ron Culp

Top left: **Two MV-22B Ospreys share the deck of a Marine assault ship with a pair of AH-1W Super Cobra attack choppers and a UH-1N during OPEVAL in 2000.** Ron Culp

Top right: **After setting down on the Amphibious Transport dock USS *Duluth* (LPD-6), the deck crew hasten to chain the aircraft down.** Ron Culp

Left: **Marines practice an assault on an offshore rig using fastrope infiltration from a hovering V-22.** Ron Culp

Bottom left: **A trio of MV-22s operates from the USS *Essex* (LHD-2) during OPEVAL in 2000.** Ron Culp

Photographs on the Facing page:

Top: **Two V-22s pass an amphibious assault ship. Aircraft 14 is seen from the scanning window in the forward escape hatch.** Ron Culp

Bottom: **Two MV-22Bs in the company of a Bell UH-1N and a pair of Boeing CH-46 'Frogs'.** Ron Culp

Above: **A V-22 nears completion at Bell Helicopter Textron's Amarillo, Texas final assembly facility. In 2001 a number of completed or nearly completed V-22s were stored in Amarillo. Most were awaiting upgrades to make them safe for operational deployment.** Jay Miller

Right: **V-22 cockpit as seen from the aircraft's jump seat. The view emphasizes the close quarters of the V-22. Note the typical flight displays on the MFDs and the NVG/HUD worn by the right-seat pilot.** Boeing

Photographs on the preceding page:

Top: **The CV-22B radar testbed, aircraft 7, captured during a training flight near Edwards AFB over California's high desert.** AFFTC

Bottom: **the V-22 in its natural habitat – the deck of a Marine amphibious assault ship. Note the slight forward tilt of the nacelles and angle on the nose wheels. The V-22's ability to taxi about the deck is a great boon compared with helicopters that must be towed.** Author's collection

Above: **One of the first MV-22Bs conducts operations from the USS *Tortuga* (LSD-49) helo deck on September 10, 1999 while its proprotor blades make contrails in the humid air.** Paul Shank

Left: **The most common view Marines have of the V-22 is the gray cabin interior, although usually appearing more worn than in this new MV-22B, aircraft 13. Troops are following the USAF flight engineer aboard via the loading ramp.** DoD

Below: **Engine access panels can also serve as work stands during maintenance.** Jay Miller

Above: **Not the most attractive aircraft in the sky, but it is certainly the most unique!** Bell Helicopter

Below: **Aircraft 22 was delivered with an overall silver finish for reflectivity testing with the new lightweight paint. The photo was taken at Bell's Amarillo, Texas final assembly facility.** Bell Helicopter

Above: **One of the EMD test aircraft flies in apparent serenity over a Maryland coastline near Patuxent River NAS.** NAVAIR

Below: **The first EMD test aircraft, ship 7, cruises in airplane mode above wetlands during flight testing in the latter half of the 1990s.** NAVAIR

Above: **The Model 609 during the course of its second flight on March 11, 2003. The crew (see inset, below) consisted of pilot Roy Hopkins (rt.) and co-pilot Dwayne Williams (lt.).** Jay Miller

Below: **The Model 609 during post-flight maintenance in the main Bell flight test facility hangar. Note that the engine nacelles are rotated to near horizontal flight attitude for maintenance access.** Jay Miller